BOTANICAL SINGAPORE
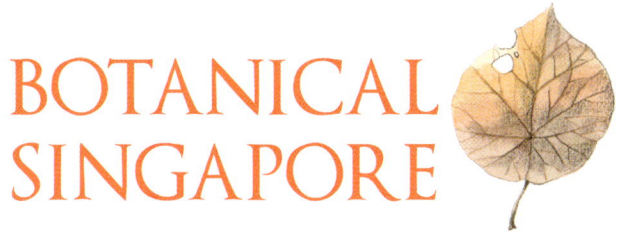

BOTANICAL SINGAPORE

An illustrated guide to popular plants and flowers

William Sim

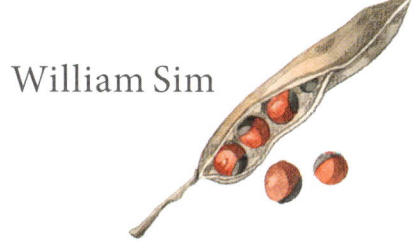

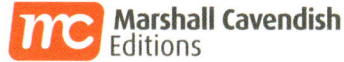

All illustrations by William Sim

© William Sim & Marshall Cavendish International (Asia) Private Limited
Published in 2017, reprinted 2022, by Marshall Cavendish Editions
An imprint of Marshall Cavendish International
1 New Industrial Road, Singapore 536196

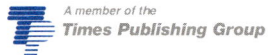

All rights reserved.

No part of this publication may be reproduced, stored in a retrieval system or transmitted, in any form or by any means, electronic, mechanical, photocopying, recording or otherwise, without the prior permission of the copyright owner. Request for permission should be addressed to the Publisher, Marshall Cavendish International (Asia) Private Limited, 1 New Industrial Road, Singapore 536196. Tel: (65) 6213 9300. E-mail: genref@sg.marshallcavendish.com.
Website: www.marshallcavendish.com/genref

The publisher makes no representation or warranties with respect to the contents of this book, and specifically disclaims any implied warranties or merchantability or fitness for any particular purpose, and shall in no event be liable for any loss of profit or any other commercial damage, including but not limited to special, incidental, consequential, or other damages.

Other Marshall Cavendish Offices:
Marshall Cavendish Corporation, 800 Westchester Ave, Suite N-641, Rye Brook, NY 10573, USA • Marshall Cavendish International (Thailand) Co Ltd, 253 Asoke, 16th Floor, Sukhumvit 21 Road, Klongtoey Nua, Wattana, Bangkok 10110, Thailand • Marshall Cavendish (Malaysia) Sdn Bhd, Times Subang, Lot 46, Subang Hi-Tech Industrial Park, Batu Tiga, 40000 Shah Alam, Selangor Darul Ehsan, Malaysia

Marshall Cavendish is a registered trademark of Times Publishing Limited

National Library Board, Singapore Cataloguing in Publication Data

Names: Sim, William, 1967- illustrator.
Title: Botanical Singapore : an illustrated guide to popular
plants and flowers / William Sim.
Description: Singapore : Marshall Cavendish Editions, [2016]
Identifiers: OCN 958358635 | 978-981-4751-96-4 (hardcover)
Subjects: LCSH: Plants—Singapore—Pictorial works. |
Flowers—Singapore—Pictorial works.
Classification: DDC 581.95957—dc23

Printed in Singapore

Contents

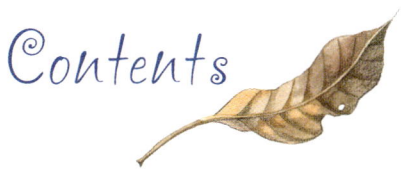

Introduction	7
Peacock Flower	8
Butterfly Pea	10
Seedsucker	12
Candelabra	14
Javanese Ixora	16
Heliconia	18
Dracaena Fragrans	20
Spider Lily	22
Sendudok	24
Red-Leaf Hibiscus	26
Mimosa	28
Caladium Bicolour	30
Ridley's Staghorn Fern	32
Bougainvillea	34
Water Hyacinth	36
Bamboo Orchid	38
Koster's Curse	40
Croton	42
Golden Shrimp Plant	44

Contents

Lantana	46
Coleus Blumei	48
Madagascar Periwinkle	50
Canna Tropicanna Gold	52
Money Plant	54
Cattleya	56
Dicranopteris Linearis	58
Episcia	60
Morning Glory	62
Raffles Pitcher Plant	64
Tillandsia	66
Pomegranate	68
Vriesea Splendens	70
Desert Rose	72
Ficus Benjamina	74
Calathea Medallion	76
Lotus	78
About the Author	80

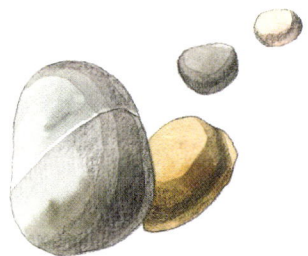

Introduction

As the Garden City, Singapore is home to over 1,300 species of plants and flowers. Our senses are held spellbound by everything from exotic blooms that burst with colour to common flora and lush greenery that beautify the country's public spaces.

Botanical Singapore pays tribute to our Garden City and award-winning artist William Sim captures some of our favourite plants in his own whimsical style. The delicate and exquisite watercolour and pencil illustrations also incorporate William's signature robots and mechanical objects.

Each illustration is accompanied by short succinct text identifying the plant and its key features, making it an attractive keepsake.

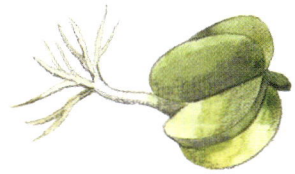

Peacock Flower
CAESALPINIA PULCHERRIMA

This ornamental shrub has yellow, red or orange flowers.
It has compound leaves that feature many tiny leaflets.
The flowers grow in a pyramidal pattern with the lower ones
having longer stalks that reduce in length for each layer above.
The Peacock Flower has medicinal properties and parts of the plant
can be used as a traditional remedy to treat various ailments.

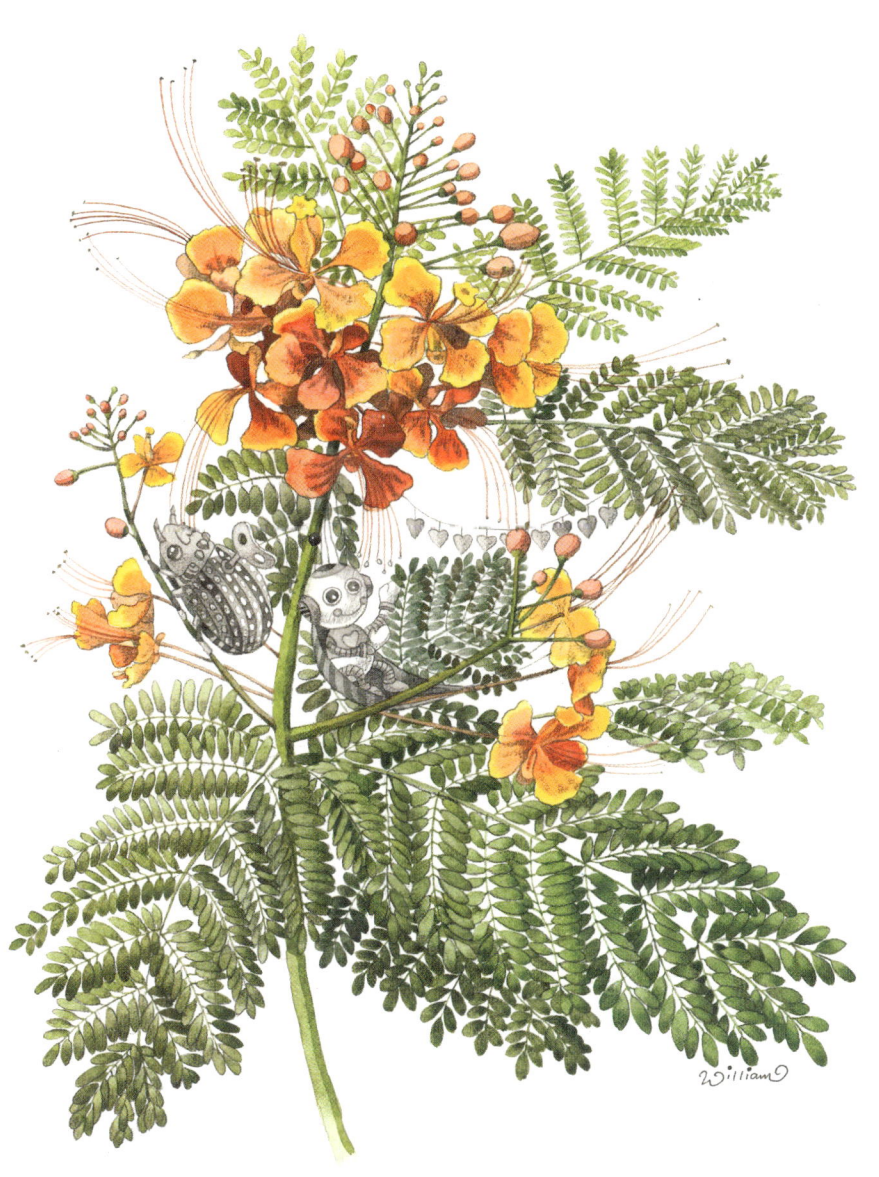

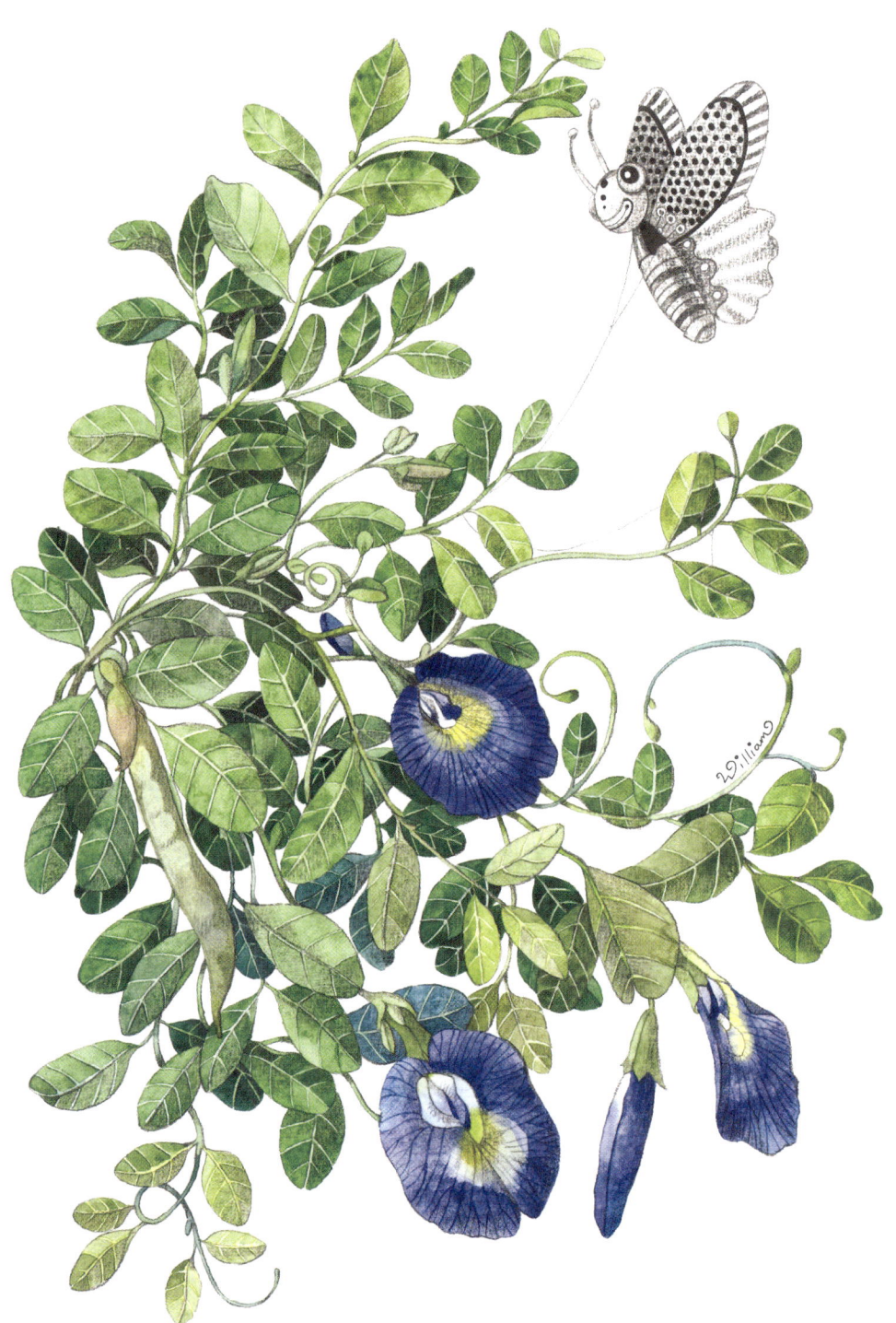

Butterfly Pea
CLITORIA TERNATEA

The Butterfly Pea is a climber plant that grows well on a trellis or fence. The characteristic blue flowers are believed to have healing properties and are used as a natural food colouring. The flowers are also used as an ingredient in a variety of teas and beauty products. The plant is very hardy and grows well in the sun.

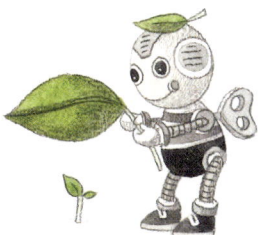

Seedsucker
GEOGENANTHUS UNDATUS

This attractive plant has leaves that appear to have dark green "stripes" on a green and purple base. It is called "Seedsucker" as the plant's leaves have a wrinkled look similar to the fabric texture of the same name. The plant is a succulent and it has thick, heavy leaves and stems that store water.

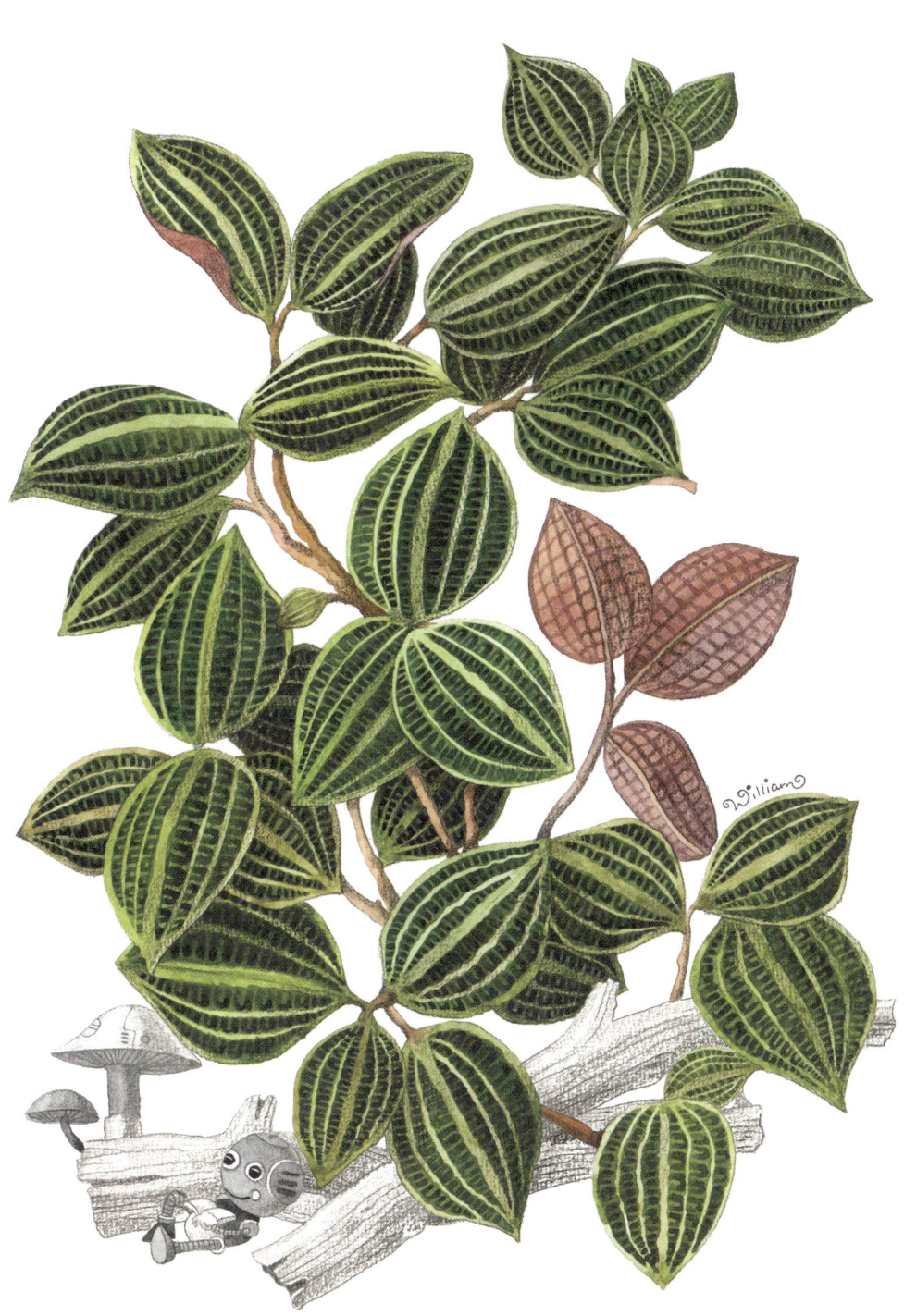

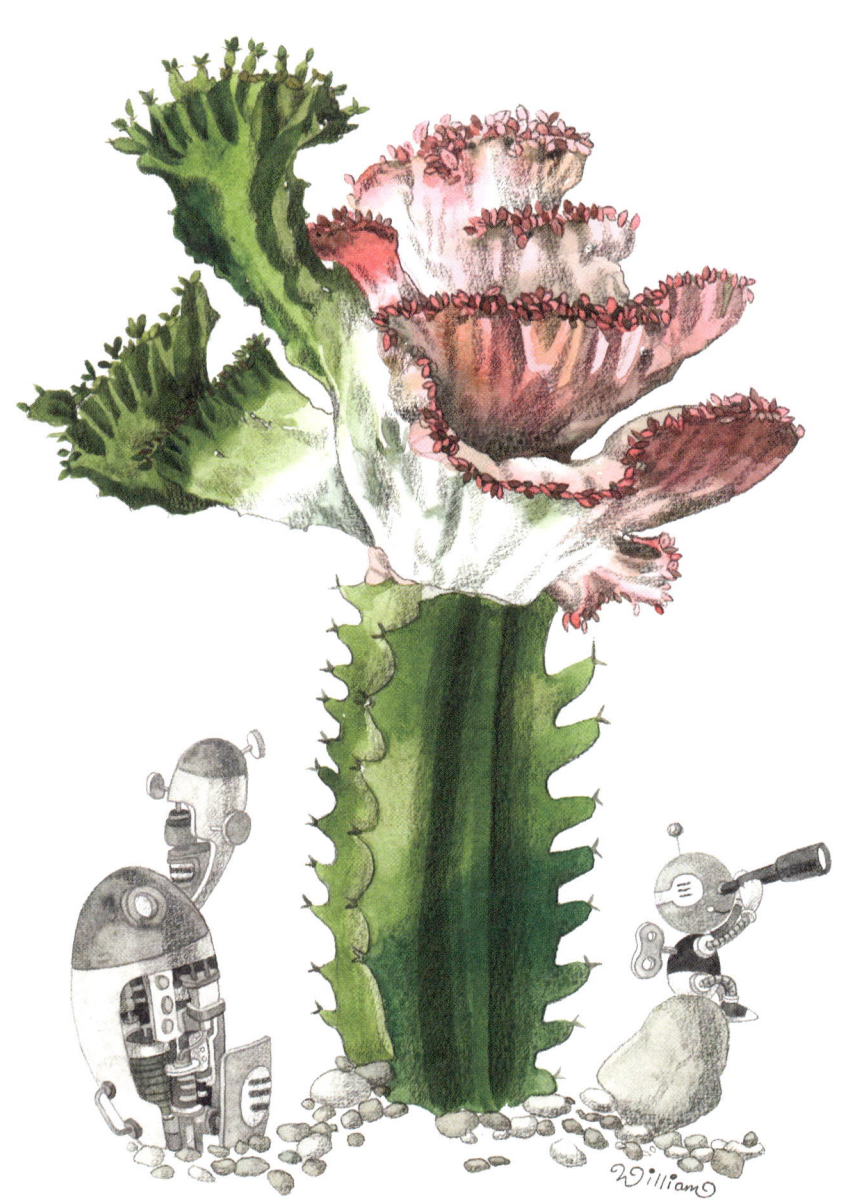

Candelabra
EUPHORBIA LACTEA

The Candelabra looks like a cactus and has three-angled branches that have scalloped edges with spines. It grows well in the sun and requires very little water. The plant seldom flowers but when it does, the flowers are small. Although the plant looks very attractive, it is poisonous and can cause diarrhoea and vomitting if ingested. The white milky sap can also cause irritation if it comes into contact with our skin.

Javanese Ixora
IXORA JAVANICA

The Ixora is a bushy, flowering shrub found in many parts of the tropics. Its flowers are usually red or pink but there are also other varieties of Ixora with yellow or white flowers. The plant has narrow, spear-shaped leaves.

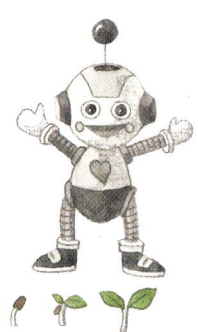

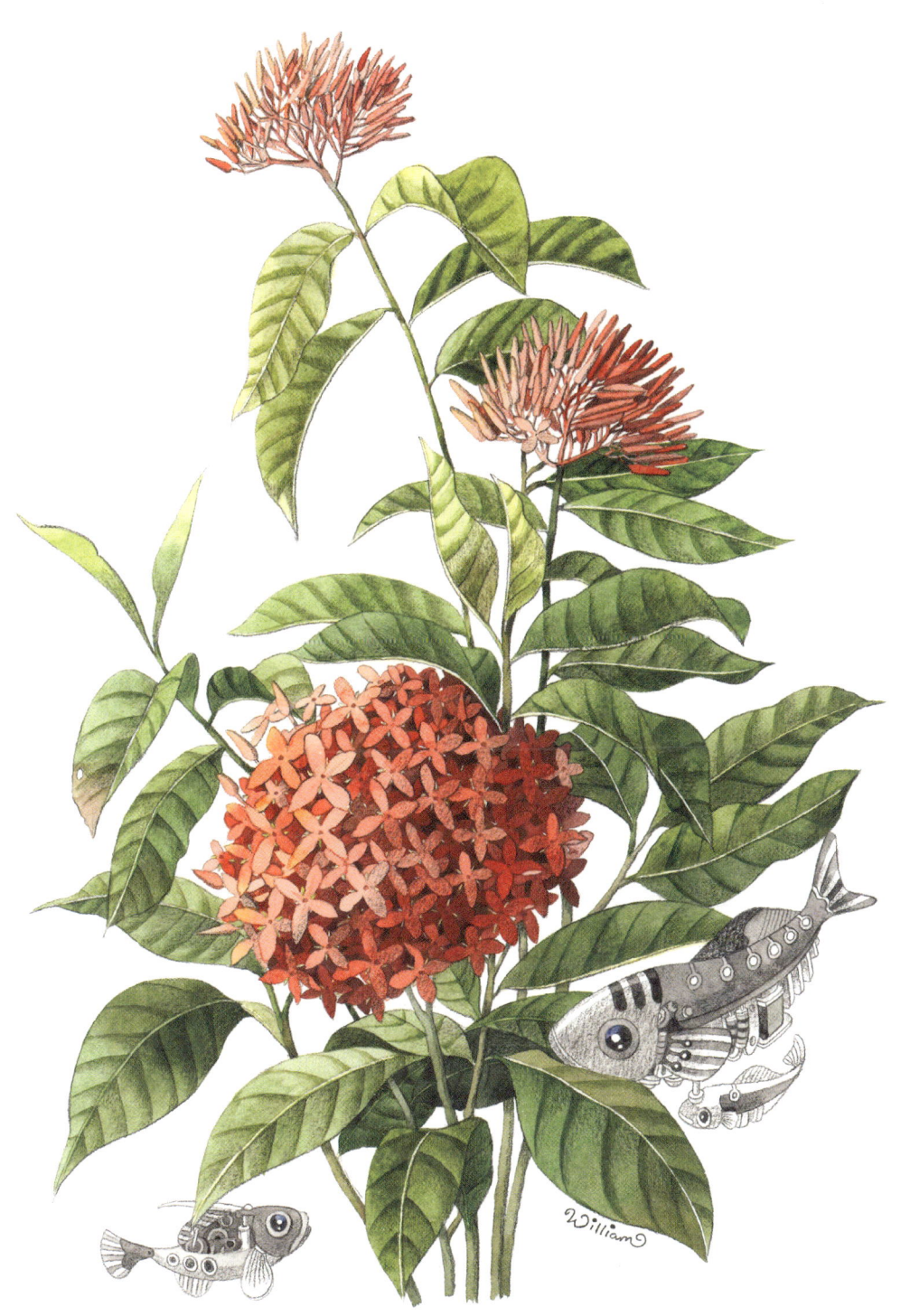

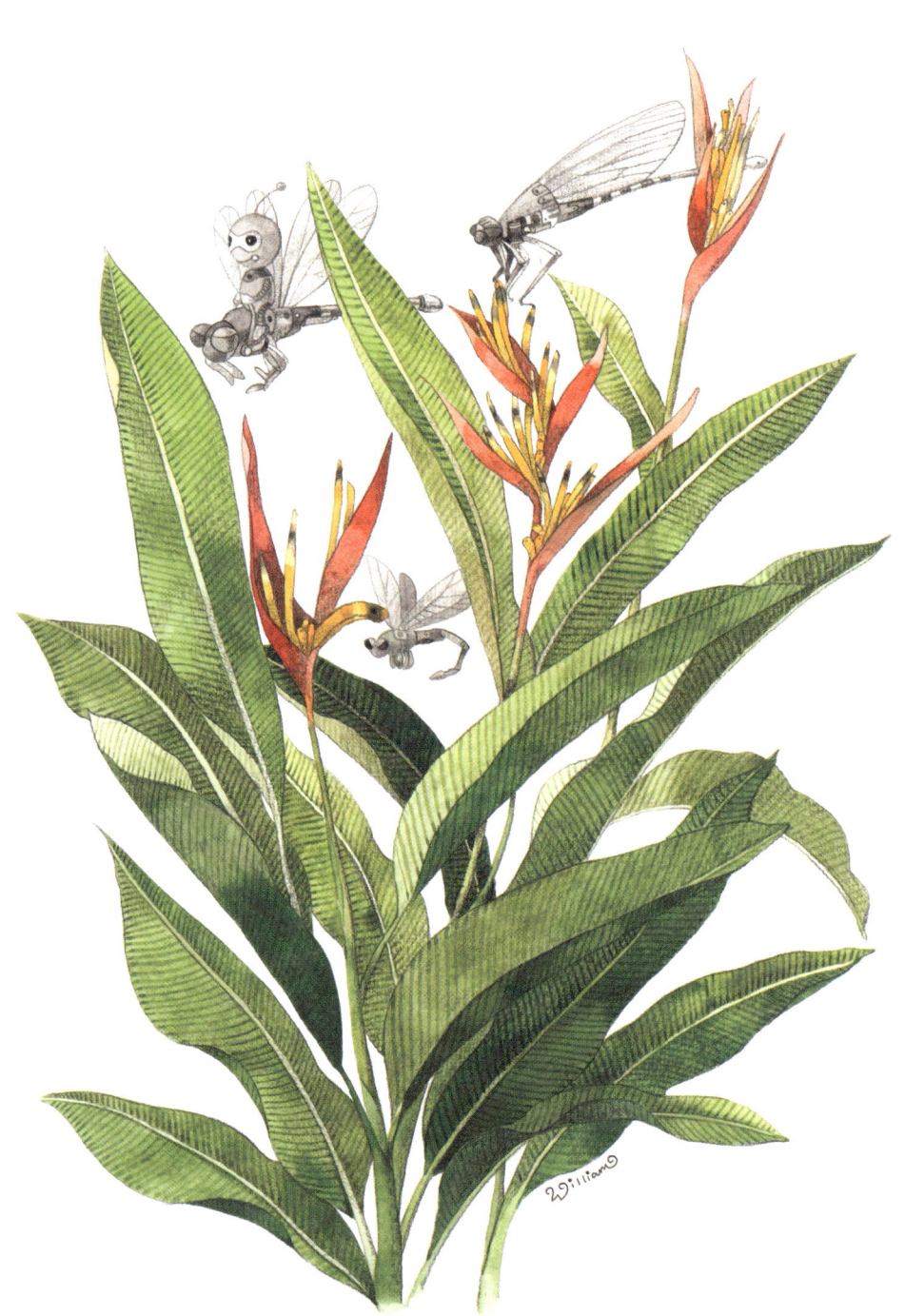

Heliconia

HELICONIA PSITTACORUM

There are over 100 species of Heliconias with many hybrids. As the Heliconia is easy to grow and has an impressive presence, the plant can be found in many tropical gardens. Heliconias are often recognised by their colourful "flowers" which are actually a group of specialized leaves called bracts. The actual flowers are located within the bracts. The leaves are similar in shape to that of the banana plant.

Dracaena Fragrans
DRACAENA FRAGRANS MASSANGEANA

Commonly known as the corn plant, the Dracaena Fragrans is often found indoors as it is tall and narrow and does not take up space. The Massangeana variety is the most popular due to its dramatic yellow pattern running down the center of its long, wide leaves. The Dracaena is hardy and can grow well in areas with low light, but it grows best near a window with filtered light.

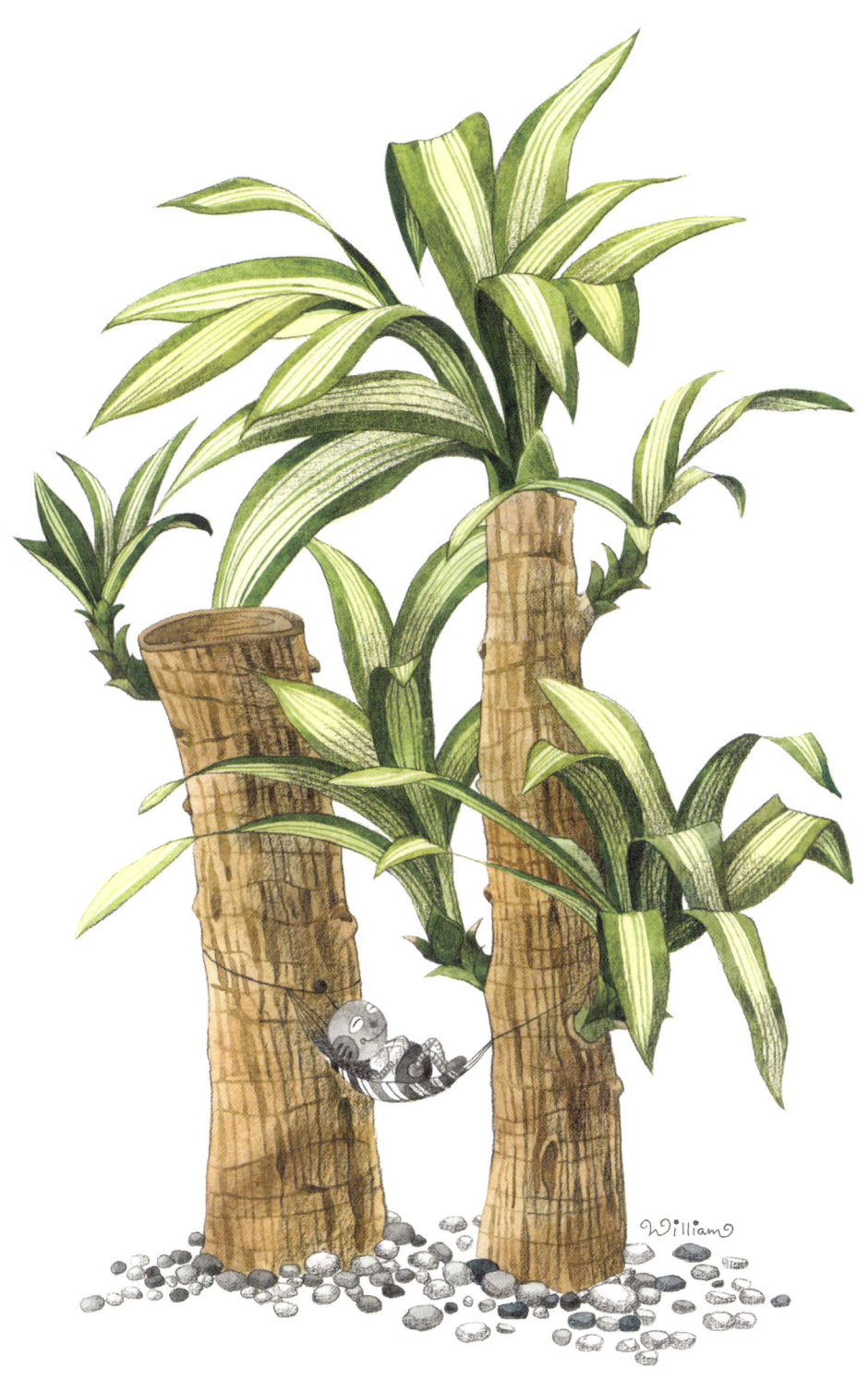

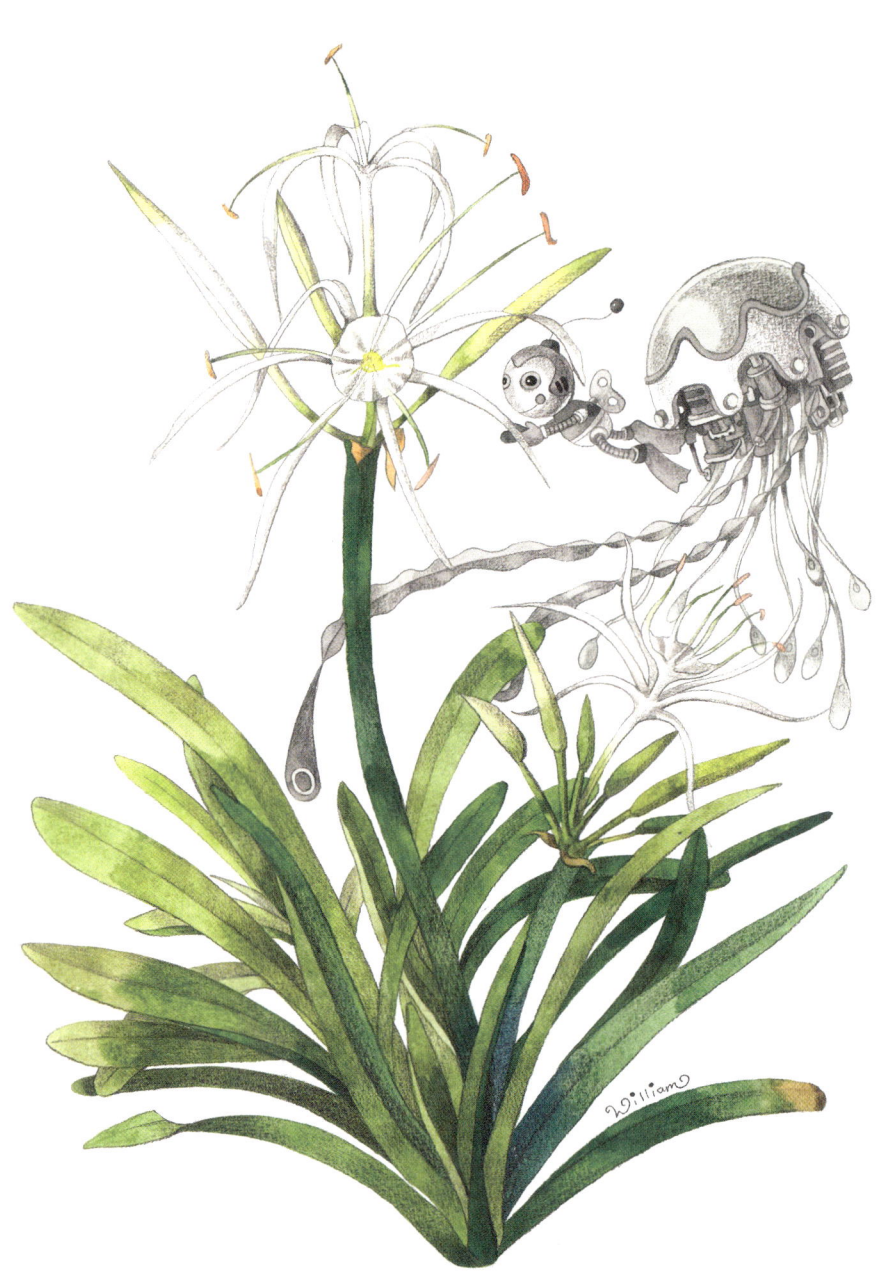

Spider Lily
AMARYLLIDACEAE

The Spider Lily is a perennial that grows easily and requires very little care. The most striking feature are the delicate white flowers which have long stamens. They are very fragrant and grow in groups of four to eight from a single stem, making them popular blooms for bouquets and garlands. The flowers also come in white, yellow, orange or red. The leaves are long and slender, about 1 cm wide and can grow up to 50 cm in length.

Sendudok
MELASTOMA MALABATHRICUM

This flowering shrub is hardy and often grows wild in open spaces. It has long narrow leaves and bears flowers and fruits throughout the year. The flowers come in various shades of purplish-pink and grow together in bunches. The fruits are berry-like and have a blackish pulp that can stain the mouth when eaten. The plant is believed to have medicinal properties and is used as such in many communities.

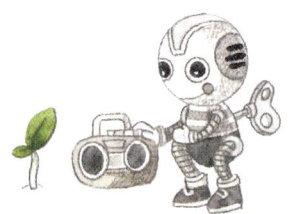

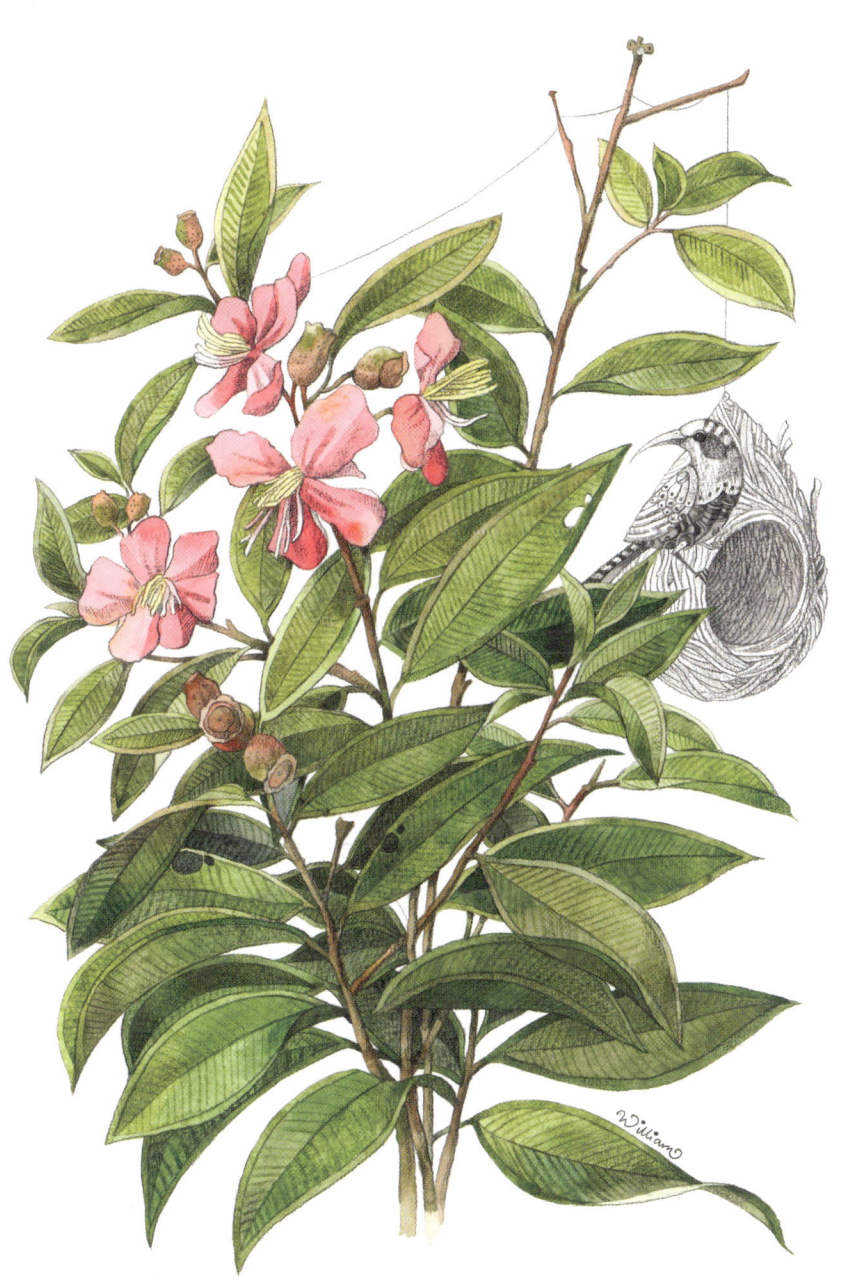

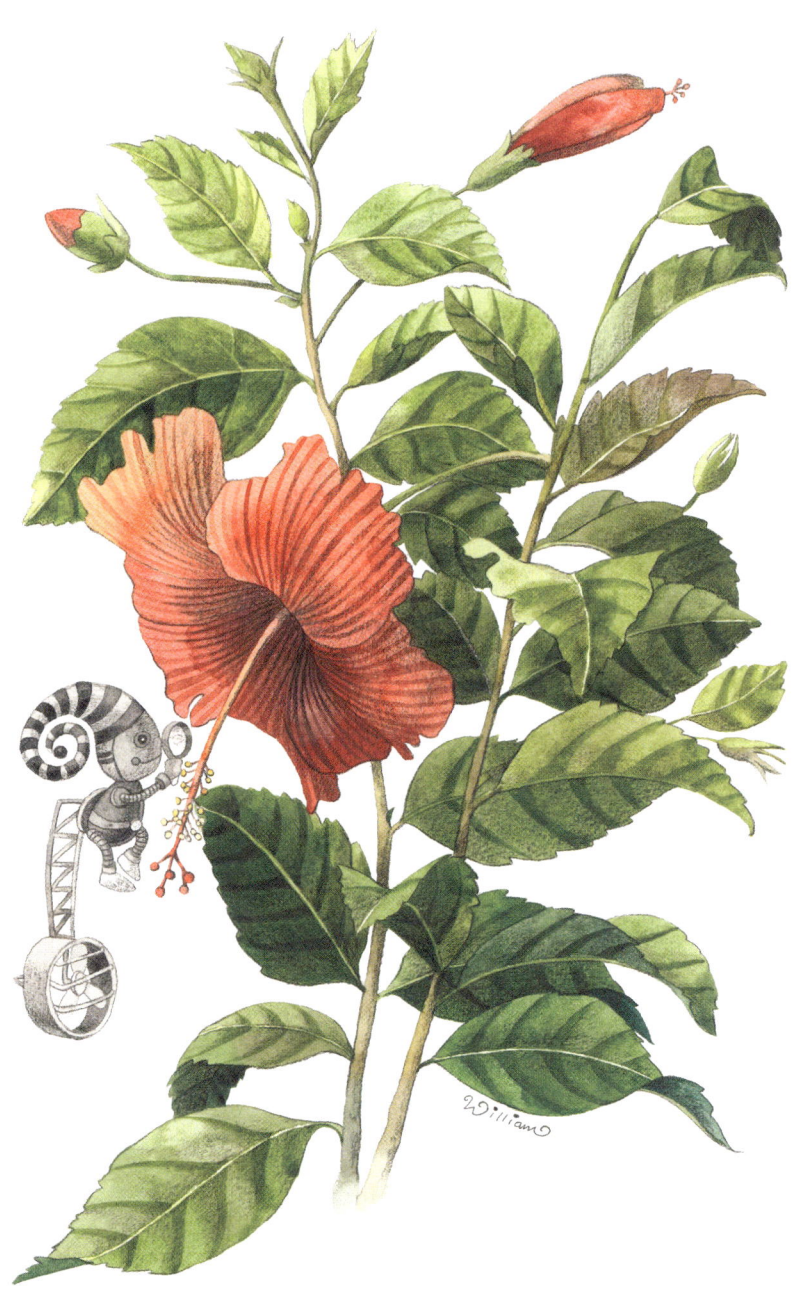

Red-Leaf Hibiscus
HIBISCUS ACETOSELLA

The flowers of this plant grow singly and the petals are joined at the base and rise from the leaf axils. The bisexual flowers can self-pollinate although cross pollination by insects does occur. The simple leaves are arranged in a spiral pattern. In Singapore, the plant grows wild to a height of 2 to 3 metres. It is often used for hedges or planted as road dividers.

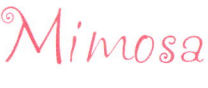

FABACEAE

This perennial shrub is easily recognised by its distinctive flowers and leaves. The pink pom-pom shaped flowers comprise hundreds of thin strand-like petals and grow in groups. The bipinnate leaves are small leaflets arranged in a feather-like format along a central spine. An interesting characteristic of the leaves is that they will fold and bend upon being touched and at night.

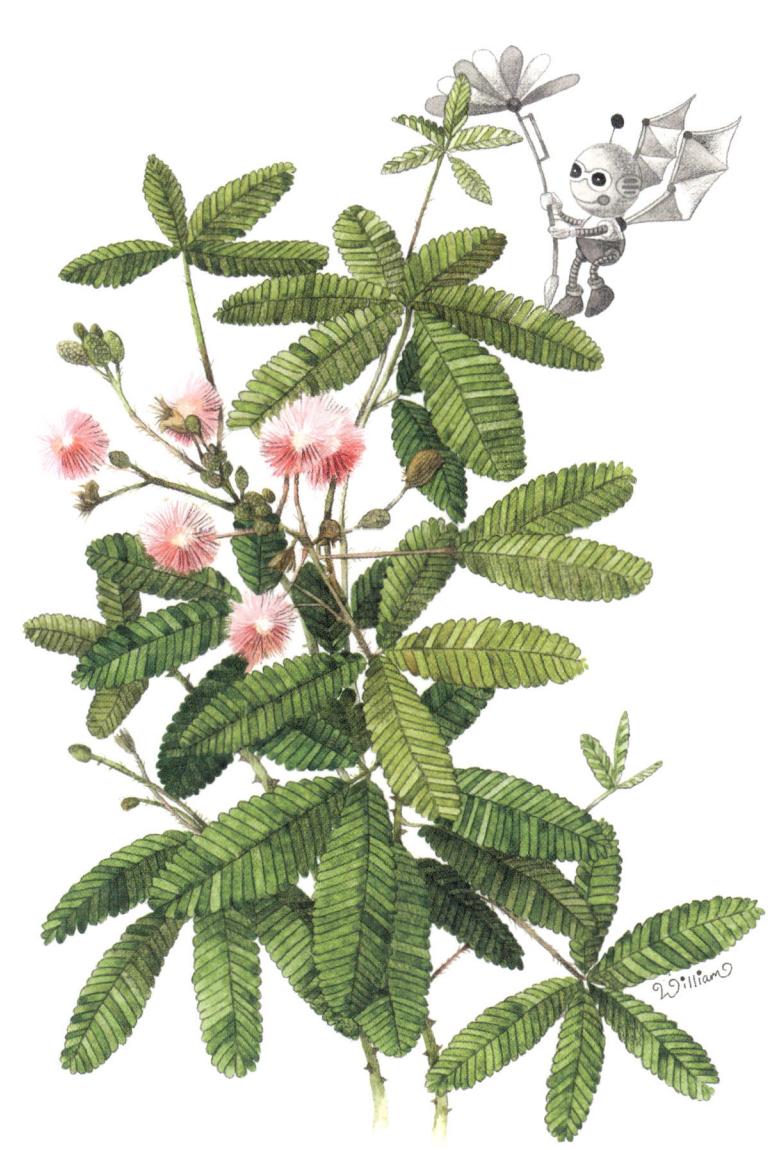

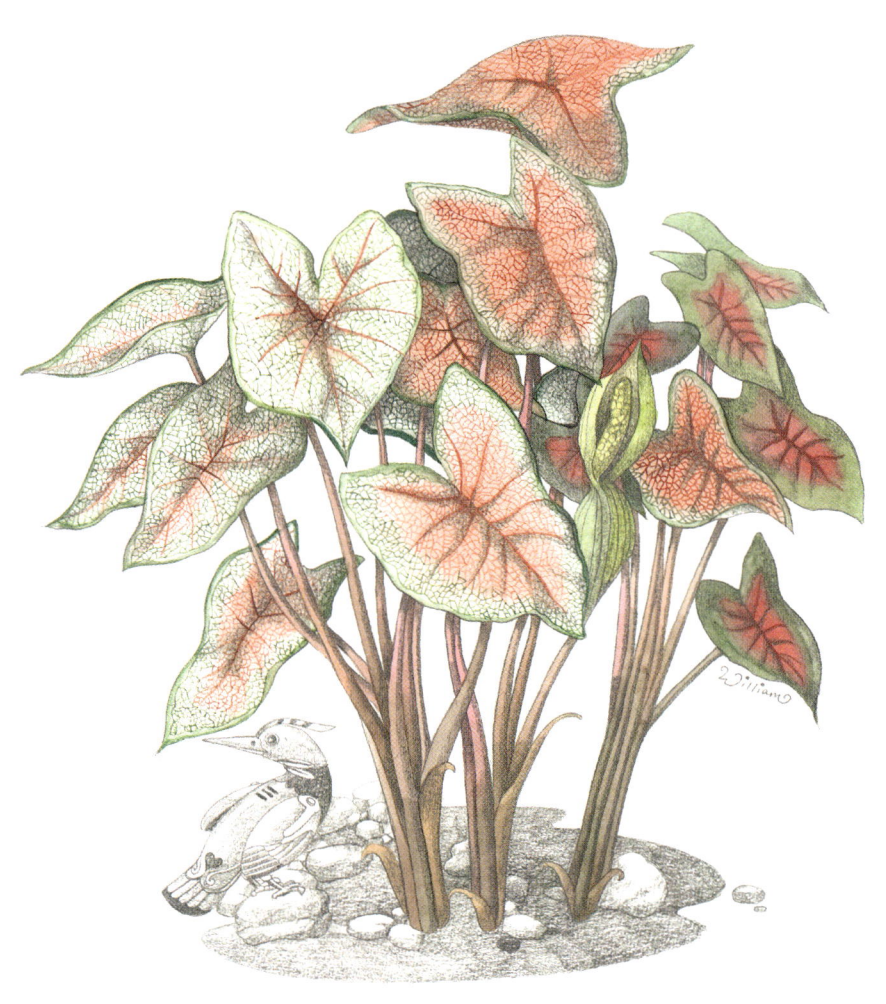

Caladium Bicolour

ARACEAE

This plant is also called elephant's ears or angel wings due to its large, thin leaves that are shaped like arrowheads. Caladiums are often used for ornamental purposes as they come in various colours and pattern combinations such as white, pink and red. Caladiums grow from tubers and can reach a height of about 60 cm. They grow best in semi-shaded or shaded areas.

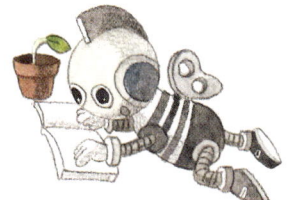

Ridley's Staghorn Fern
PLATYCERIUM RIDLEYI

These ferns are commonly called staghorn or elkhorn ferns due to their uniquely shaped fronds. They are epiphytes which are plants that grow harmlessly on another plant (such as a tree) and draws its moisture and nutrients from its environment and not its host. They used to be found growing on tall trees in the Bukit Timah Nature Reserve but are now extinct to Singapore. What we see today are sourced from nurseries overseas.

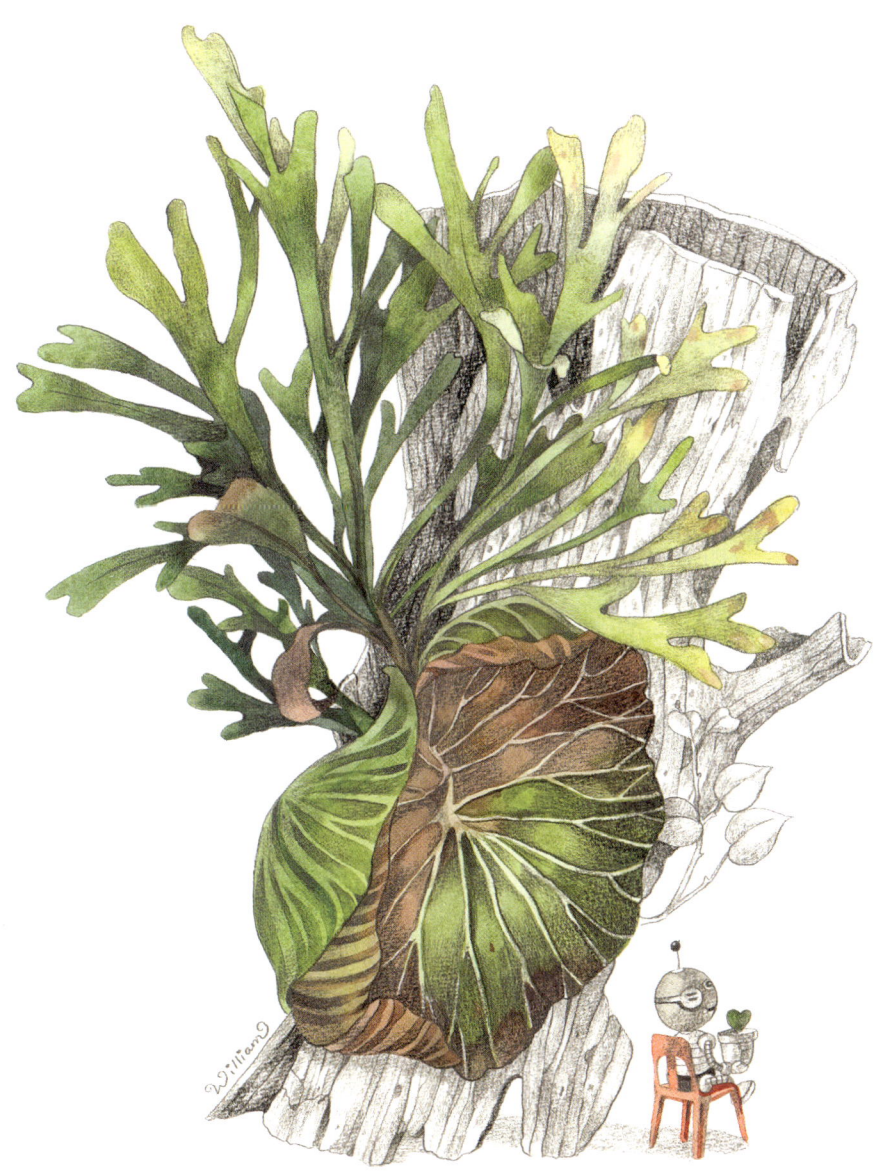

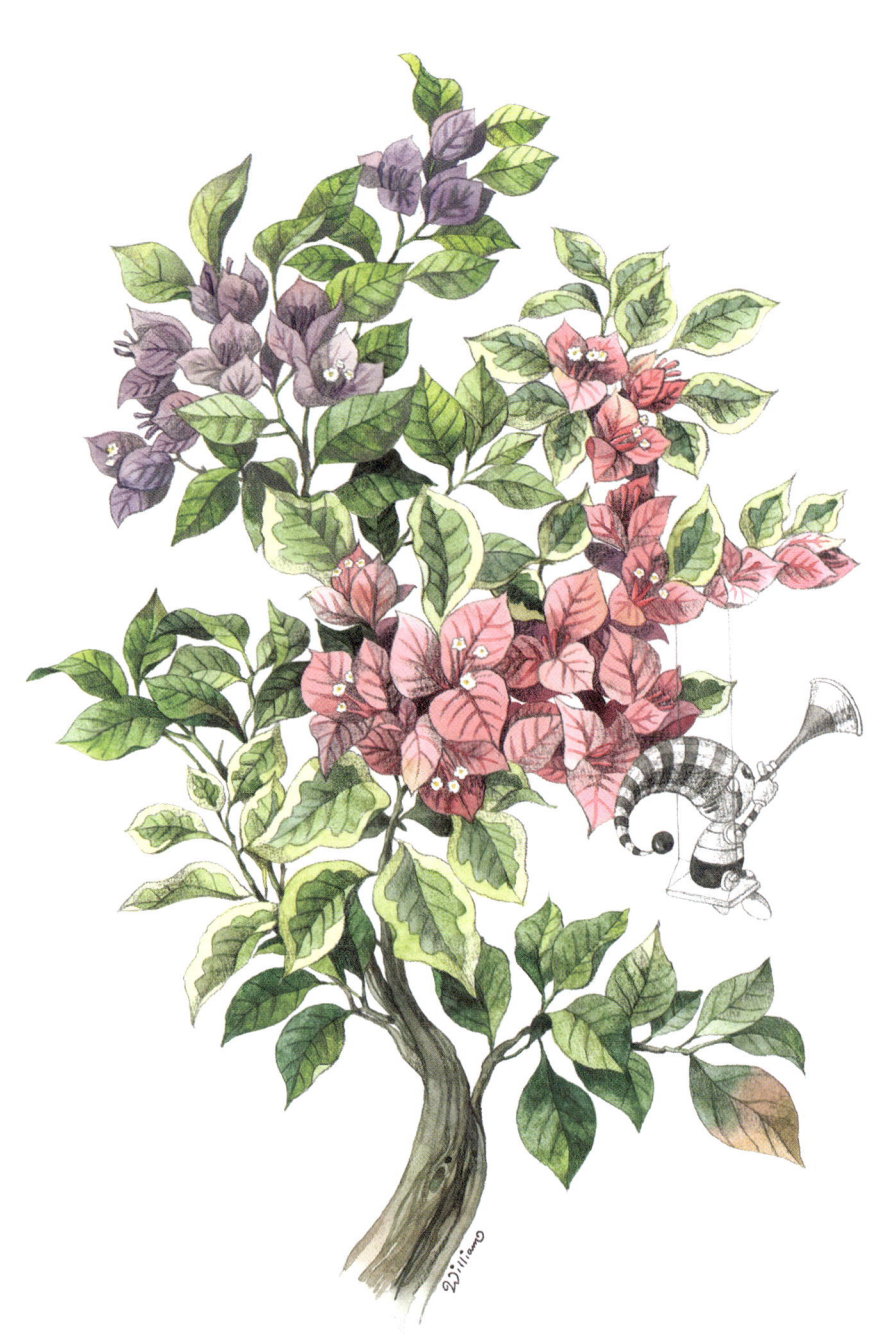

Bougainvillea
NYCTAGINACEAE

The Bougainvillea is planted in many parks and along road dividers in Singapore as it is easy to maintain; it can grow well in the sun with very little water. The flowers are actually small and white, and what many people think are its flowers are actually bracts or special leaves. The bracts come in many bright colours such as pink, magenta, purple, red, orange, white or yellow.

Water Hyacinth
EICHHORNIA CRASSIPES

An aquatic plant, the Water Hyacinth has large, roundish leaves that float above the water surface. The purplish flowers grow in groups of between 8 to 15 flowers from long spongy stalks. The fibrous roots are long and feathery. The plant grows very quickly, and can become a pest if its growth is not controlled. It covers the water body entirely, blocking sunlight which kills the native aquatic plants. It also drains oxygen from the water, killing the fish and other creatures that live there.

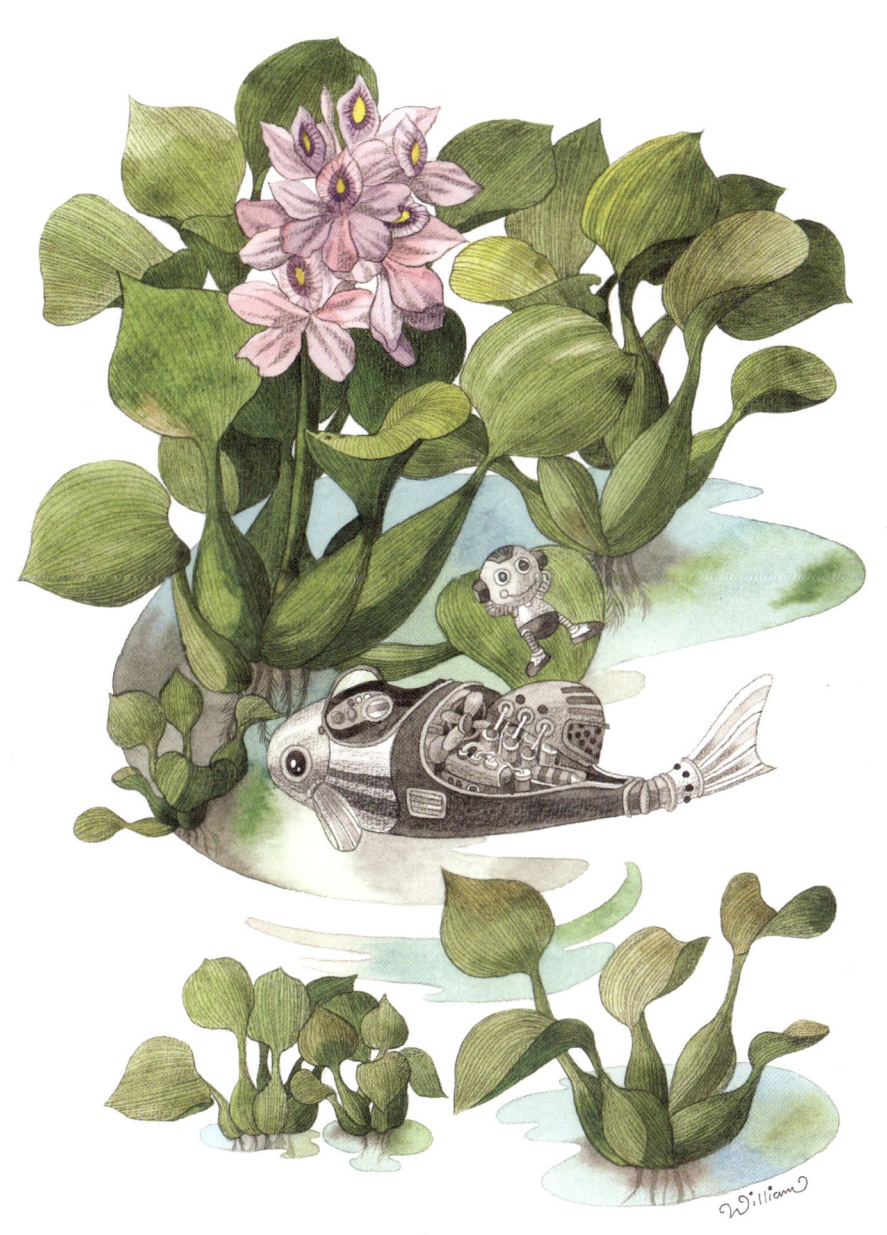

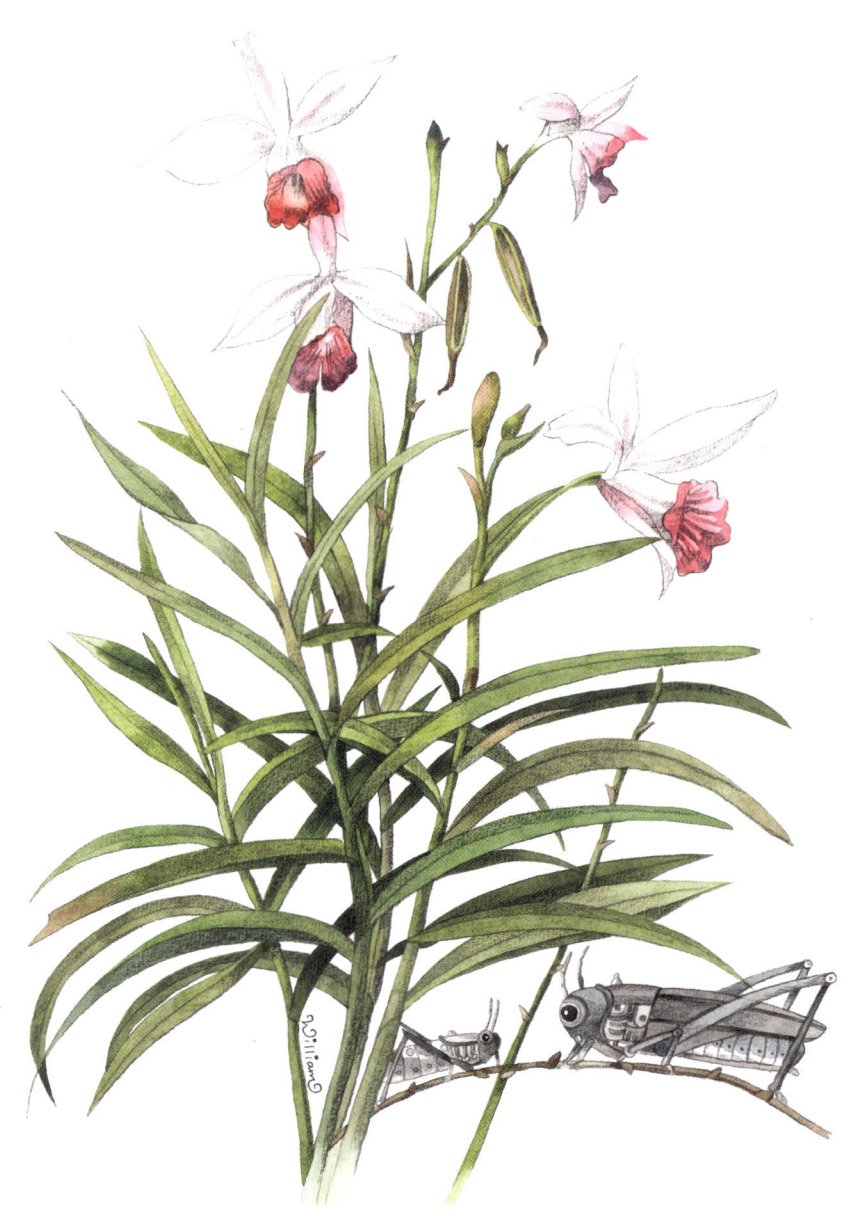

Bamboo Orchid
ARUNDINA GRAMINIFOLIA

The long grass-like leaves of this plant grow on the stems which are joined at the base. The stems can grow up to 2.5 metres tall. The large flowers grow at the tips of the stems and vary in colour, ranging from white to pink with a purple lip.

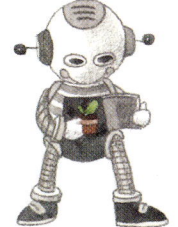

Koster's Curse
CLIDEMIA HIRTA

The most obvious feature of this plant are the dark green leaves with jagged edges. The leaves have five major veins that run from base to tip that create a grid-like pattern. Most parts of the plant are covered with fine, white hair. The small, white flowers have rounded petals. The fruit look like black berries and can contain over a hundred tiny seeds each. This is an invasive plant and if left unchecked, can grow in dense thickets that can smother plantations, pastures and native vegetation.

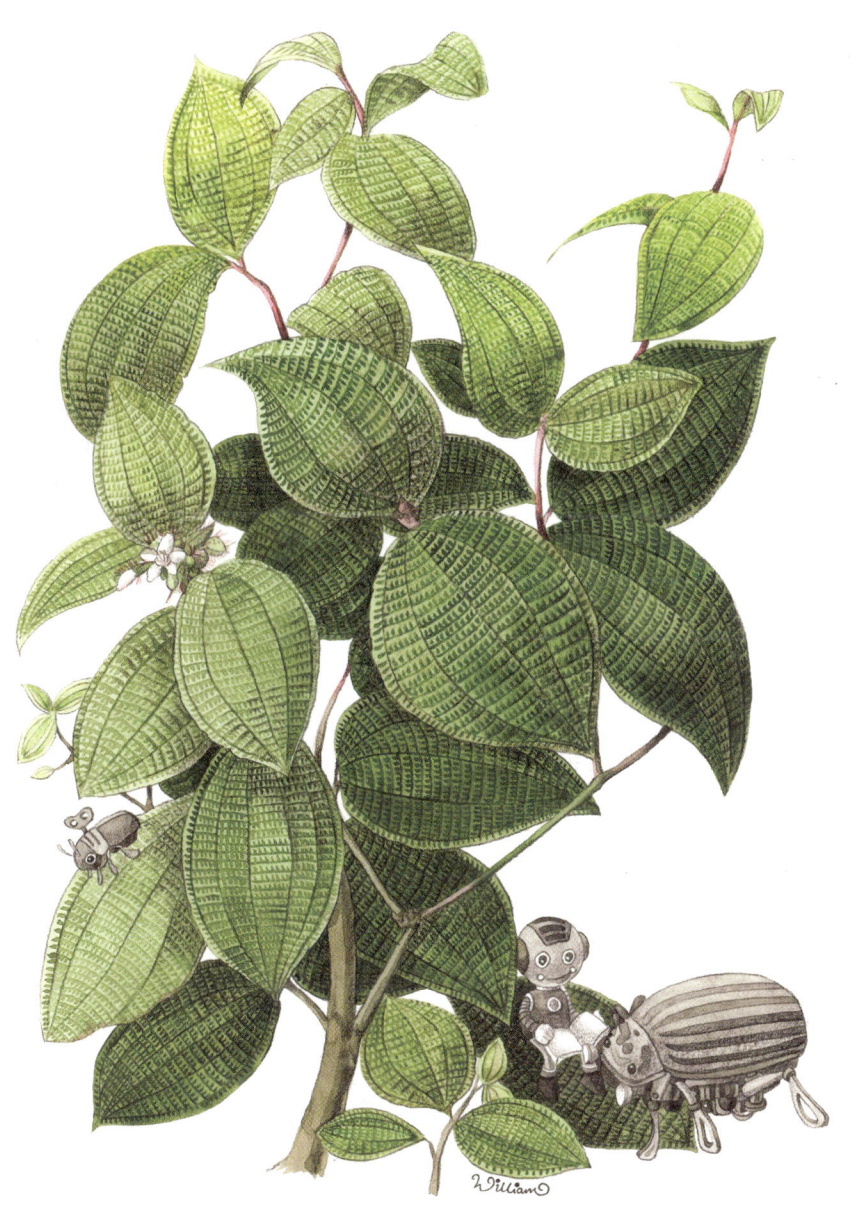

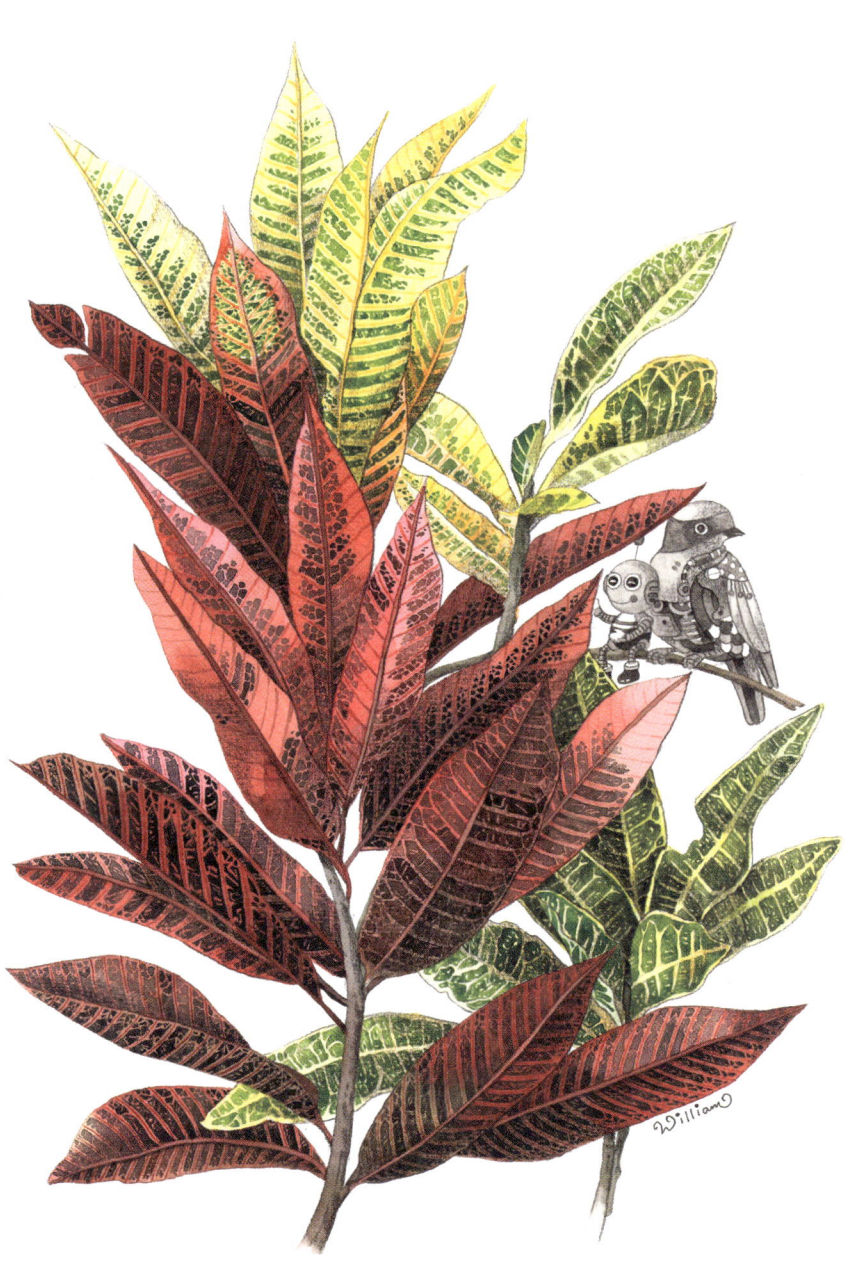

Croton

CODIAEUM VARIEGATUM

This evergreen shrub can grow to a height of 3 metres. It has smooth stems with large, thick leaves in a variety of shapes and colours such as green, yellow, purple and pink. Some species also have speckled spots or stripes on the leaves. This colourful selection makes them a popular choice for hedges and in garden landscapes. The Croton has small, inconspicious flowers.

Golden Shrimp Plant
PACHYSTACHYS LUTEA

The narrow, white tubular flowers of this plant are covered by overlapping, yellow bracts or modified leaves. These resemble the scales on a shrimp, giving the plant its name. The leaves are dark green and shaped like a lance. The contrast in colour beween the flowers and the leaves make it very suitable for use in landscaping and decorative gardens.

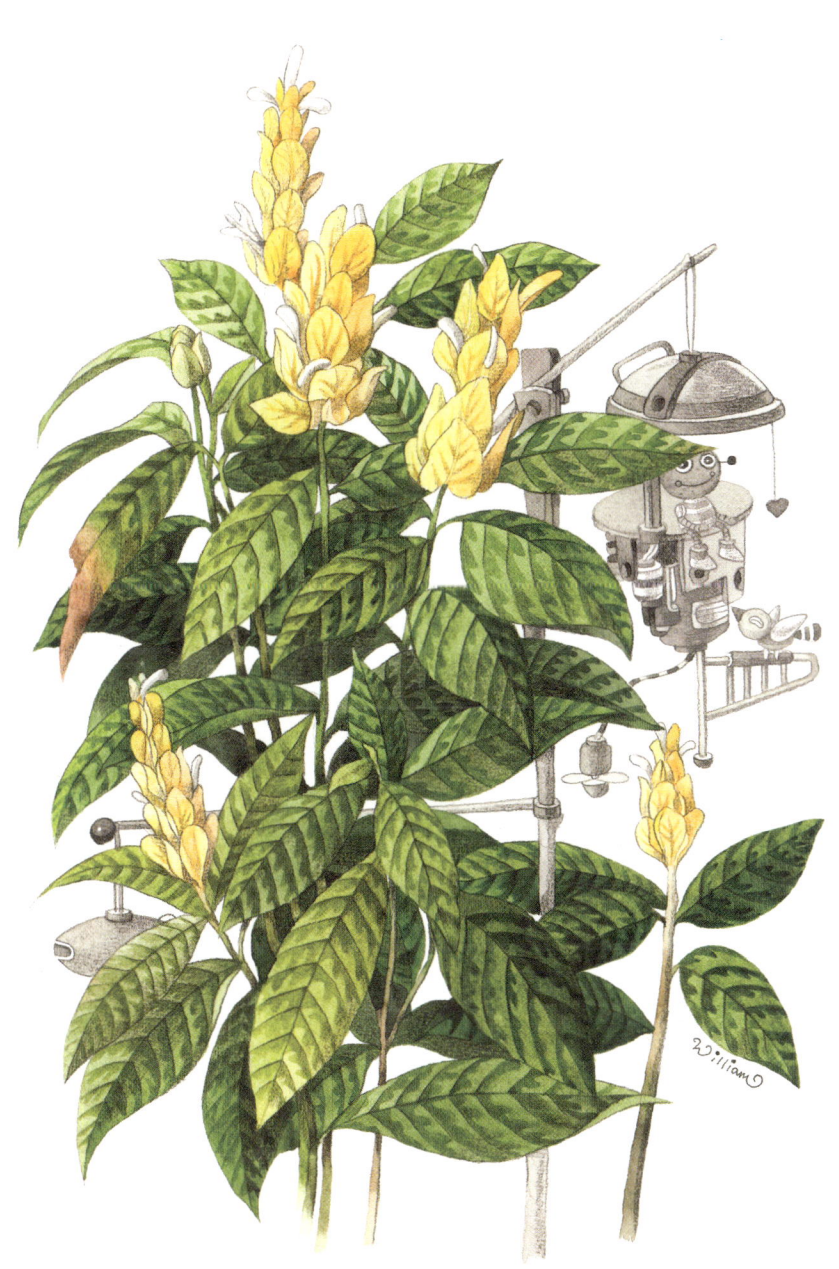

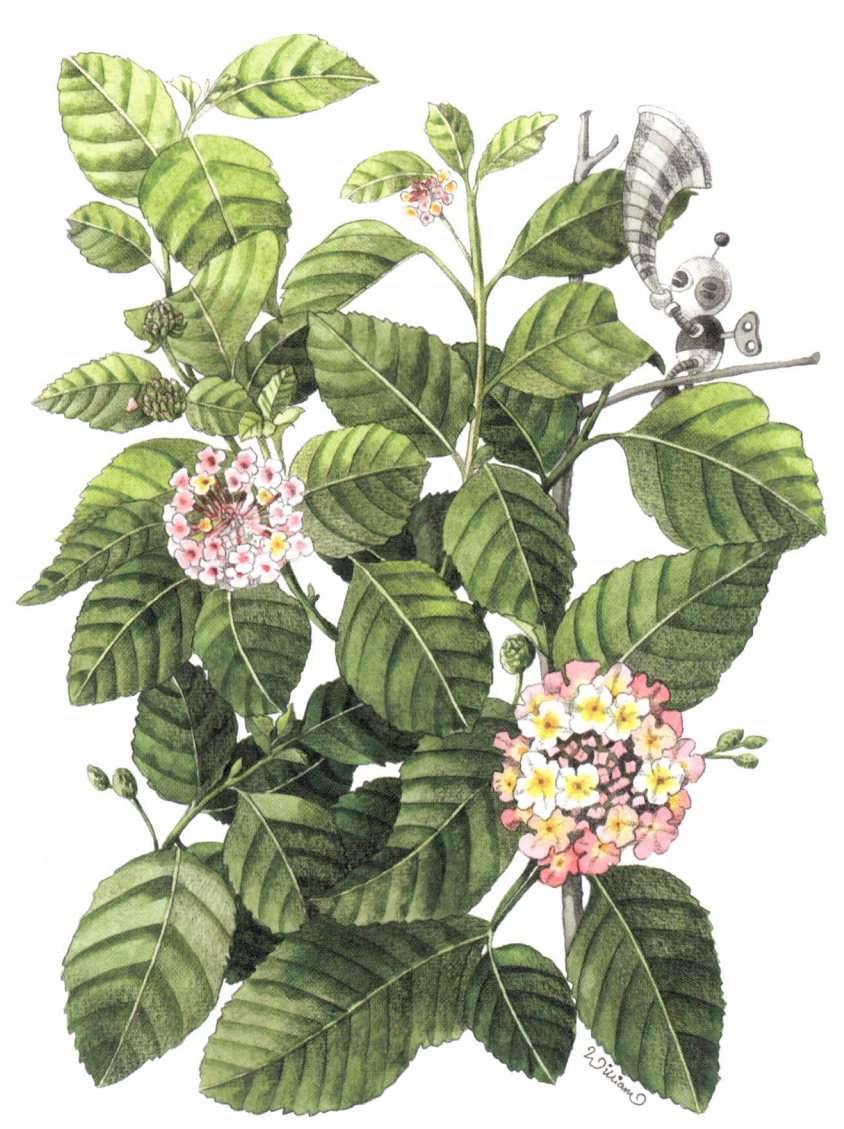

Lantana

CARYOPTERIS HISPIDA

The Lantana has aromatic flowers that grow in clusters called umbels. Although the flowers come in a mix of colours such as red, orange, yellow, blue or white, they change colour over time. It is therefore common to have two-toned or even three-toned flowers. Parts of the plant are believed to have healing properties and are used as an ingredient in traditional medicine.

Coleus Blumei
PLECTRANTHUS SCUTELLARIOIDES

The colourful leaves of this plant feature interesting patterns that come in shades of green, pink, yellow, dark purple, maroon, cream, white and red. The variation in colour contrast depends on the intensity of the sunlight it receives. The Coleus is a favourite with gardeners as it can grow outdoors and is also a low-maintenance houseplant.

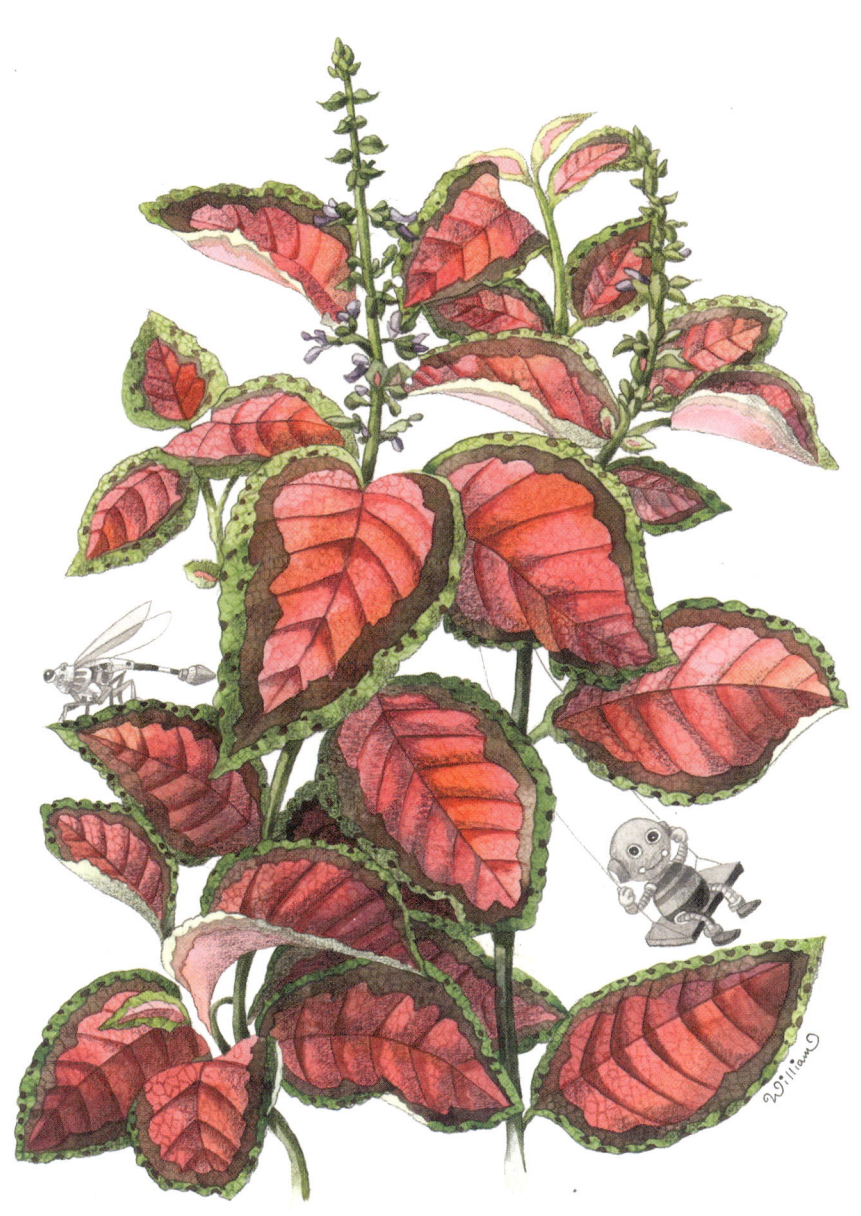

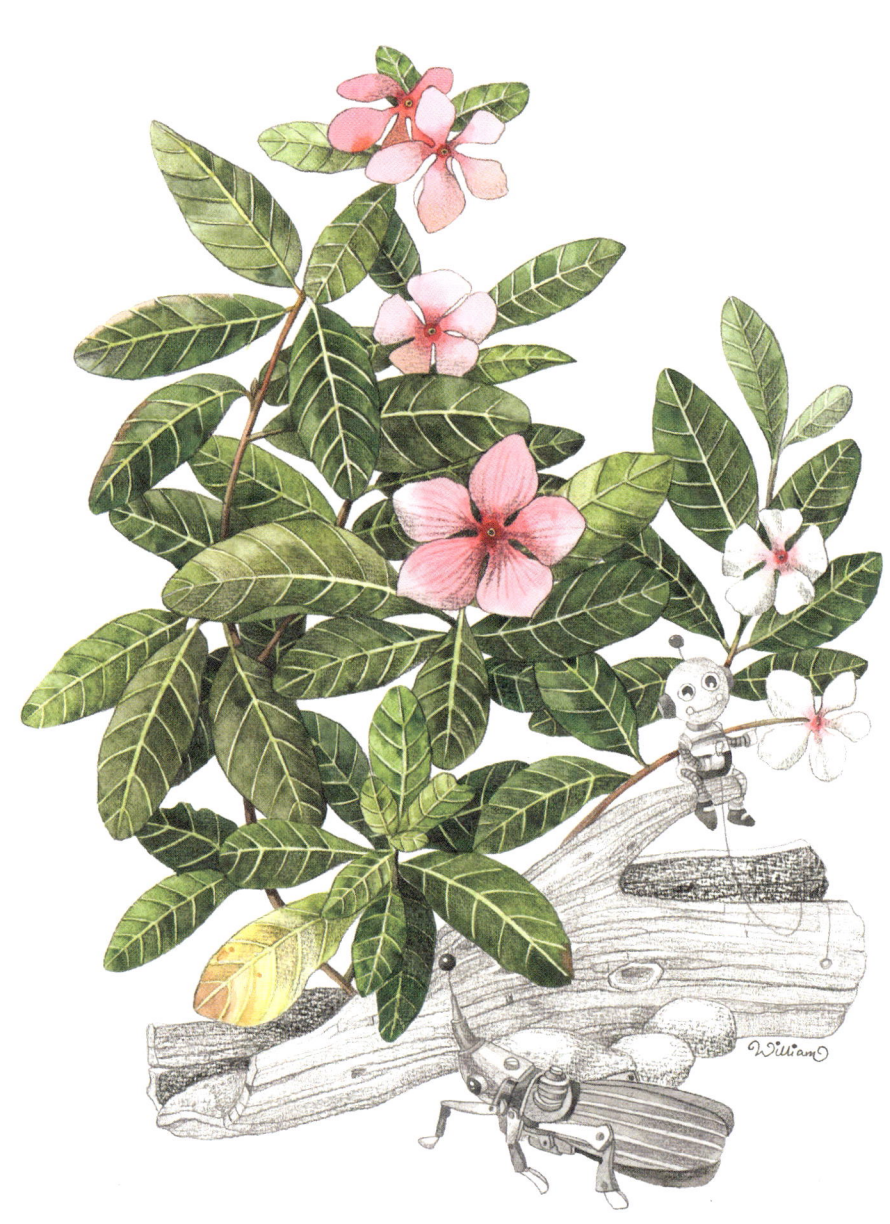

Madagascar Periwinkle
CATHARANTHUS ROSEUS

There are eight known species of Catharanthus. Seven of them are native to Madagascar, although the Roseus can be found around the world. The plant is often grown for ornamental purposes as it has attractive flowers that range in colour intensity from light pink to red. The tubular-shaped flowers grow singly and each has five petals. The dark green leaves are oval in shape.

Canna Tropicanna Gold
CANNACEAE

With its bright bold flowers and distinctive leaf patetrns,
this is a gorgeous plant that will look great in any garden.
The tall stems are crowned with large, orange flowers that have
a distinct yellow edge. The leaves have a striped pattern in
shades of green and gold that provides contrast to the flowers.
The plant is easy to maintain and grows well under direct sun.

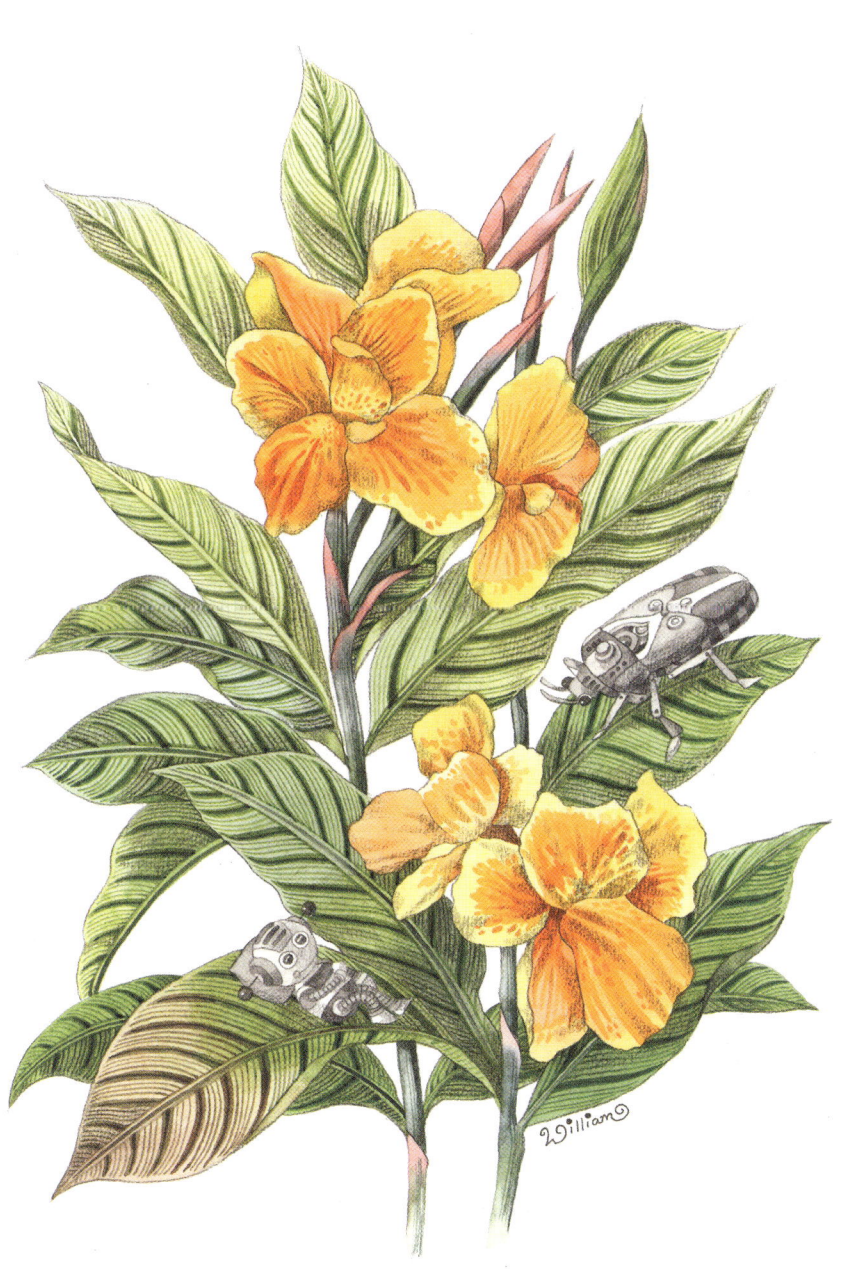

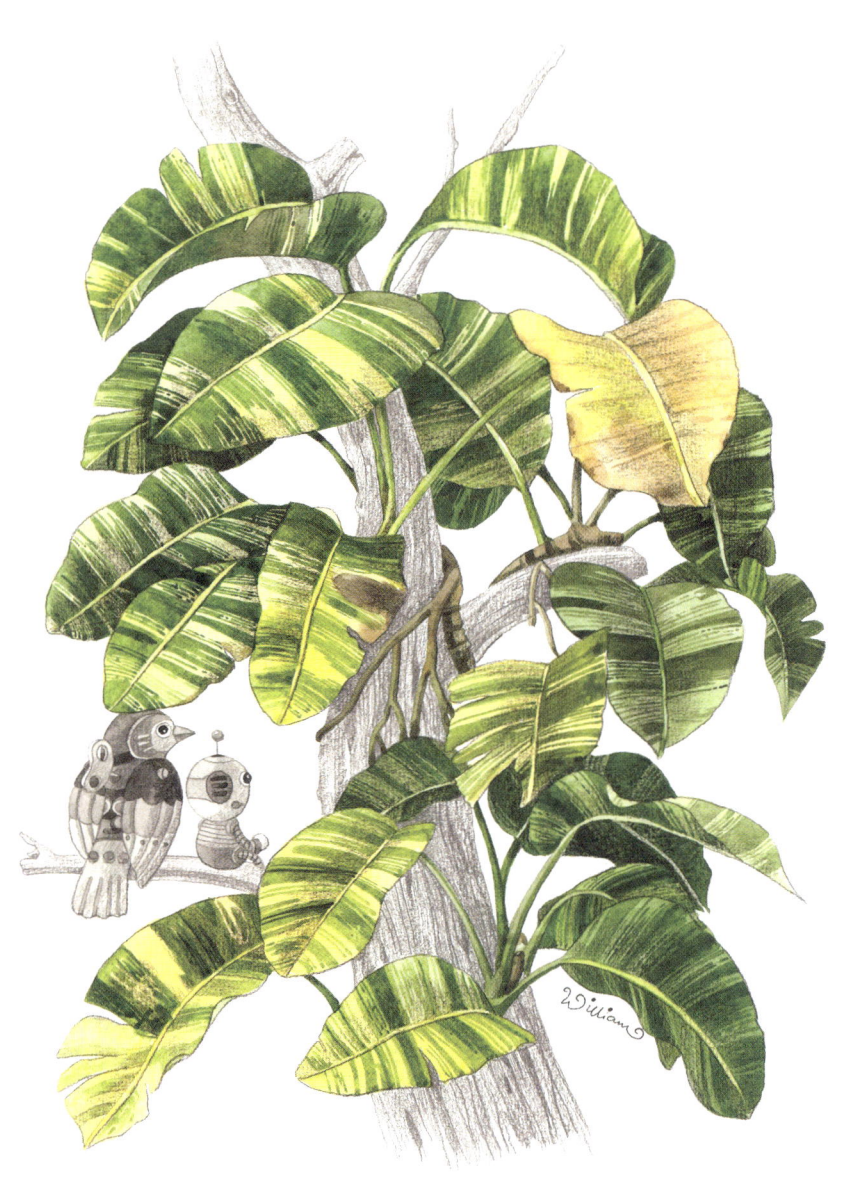

Money Plant
EPIPREMNUM AUREUM

An easy to grow indoor plant, this vine has aerial roots which allow it to climb and cling to surfaces. It has heart-shaped leaves with trailing stems that lets it "climb" trees and take root when they reach the ground. The Money Plant is often used to decorate public spaces as it is large and leafy and needs little care.

Cattleya
CATTLEYA LABIATA

The Cattleya is an orchid and as an epiphyte can grow on trees, or even on rocks where there is very little soil. Each flower has three narrow sepals and three broader petals. Two petals are similar to each other, and the third has a conspicuous lip with markings and a frilly edge. The flowers originate from a pseudobulb and each stalk can have from anything from one to a maximum of ten flowers.

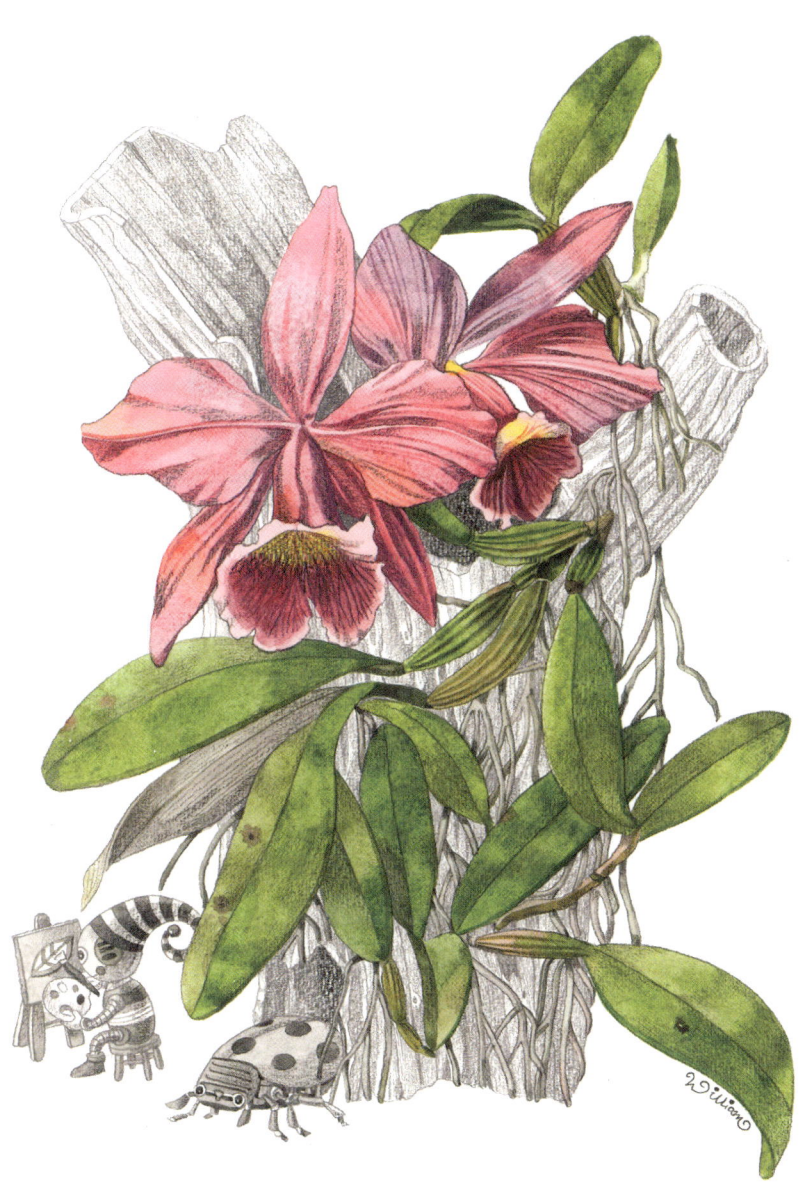

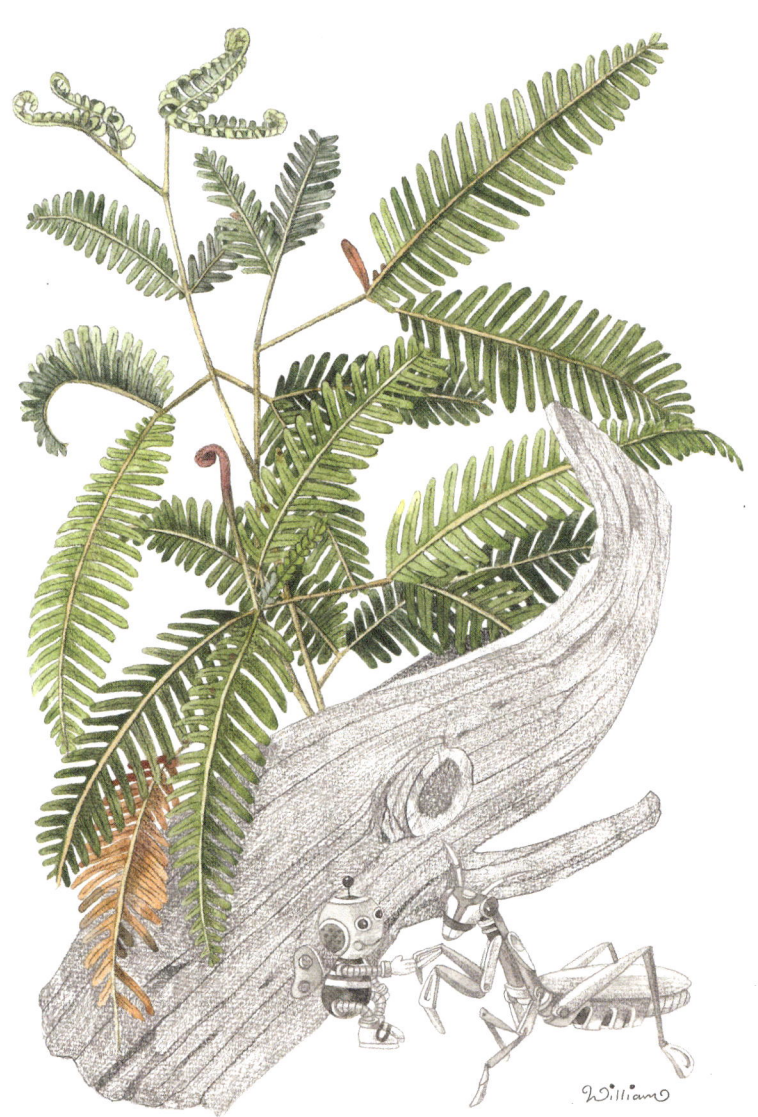

Dicranopteris Linearis
GLEICHENIACEAE

This is a large fern, growing up to 3 metres across. The plant spreads into two distinct branches with each having several compound stalk fronds. It grows easily and can even take root on lava flows or rock debris. The fern is used in traditional medicine: to treat fever in Malaysia and expel intestinal worms in Indochina.

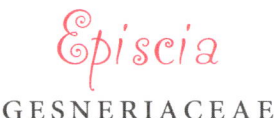

GESNERIACEAE

The Episcia has small red flowers and dark patterned leaves. It is a popular houseplant as it is attractive and flowers all year round. Its name is derived from the Greek word *episkios* meaning shady, and the plant grows best in semi-shaded areas.

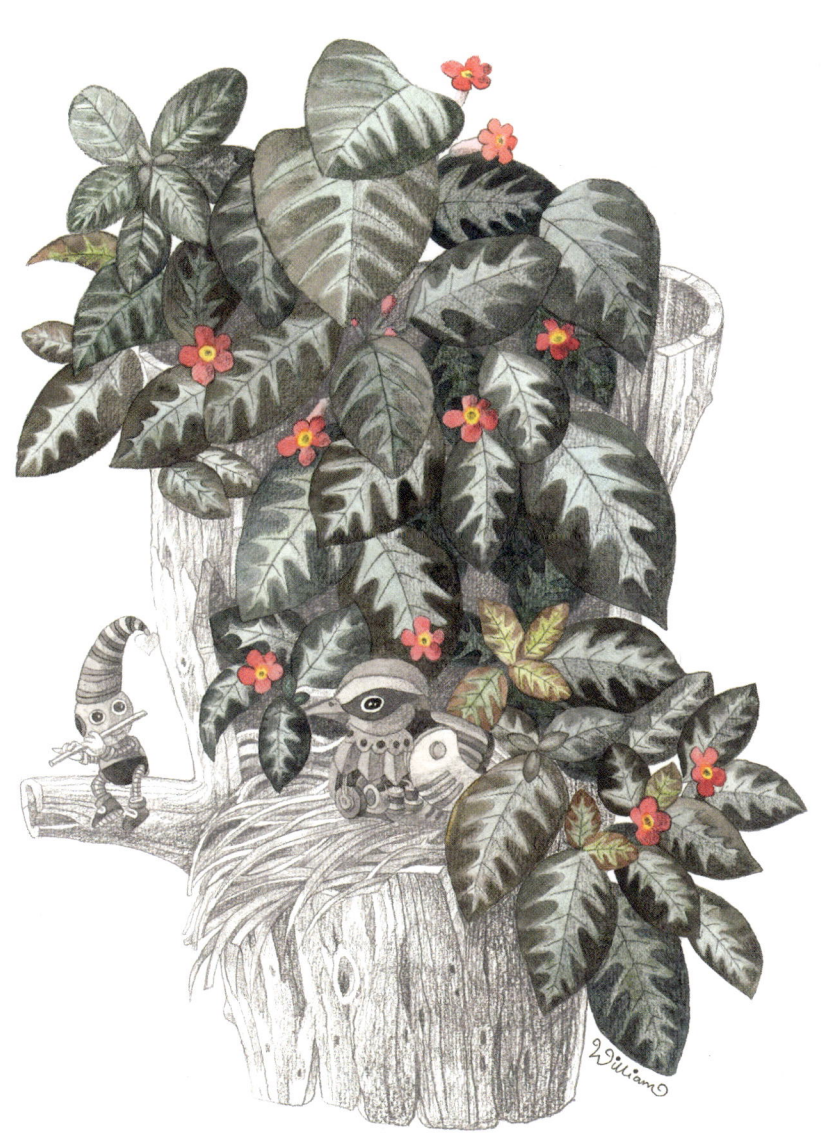

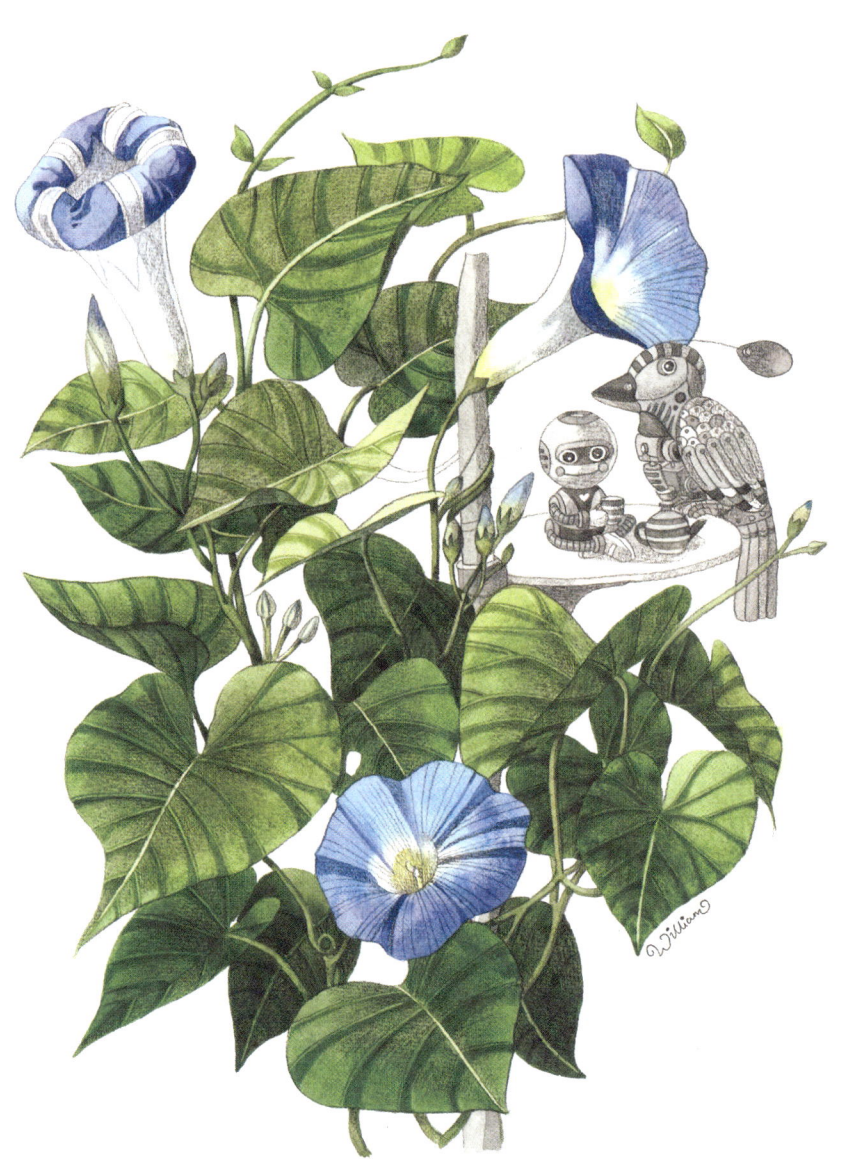

Morning Glory
IPOMOEA

Morning Glory flowers bloom fully early in the morning and fade as the day passes. The trumpet-shaped flowers range in colour from pink or blue to purple. The plant has heart-shaped leaves and slender stems. It is a climber and can be trained to grow over a trellis for landscaping purposes or to provide shade. As it bears roots easily, do plant it in the ground or in a very large pot. The flowers attract butterflies and insects that help in the pollination process.

Raffles Pitcher Plant
NEPENTHES RAFFLESIANA

This vine has long leaves and tendrils that can trail for up to 1 metre. The leaf tips terminate in coiling tendrils, which under ideal conditions become prey-catching pitchers. The pitchers are partially filled with rainwater enriched with the digestive enzymes it secretes. Their colour and size varies but the typical colour is light green throughout with heavy purple blotches on the lower pitchers and cream-coloured aerial pitchers. The plant has small yellow flowers that bloom once or twice a year.

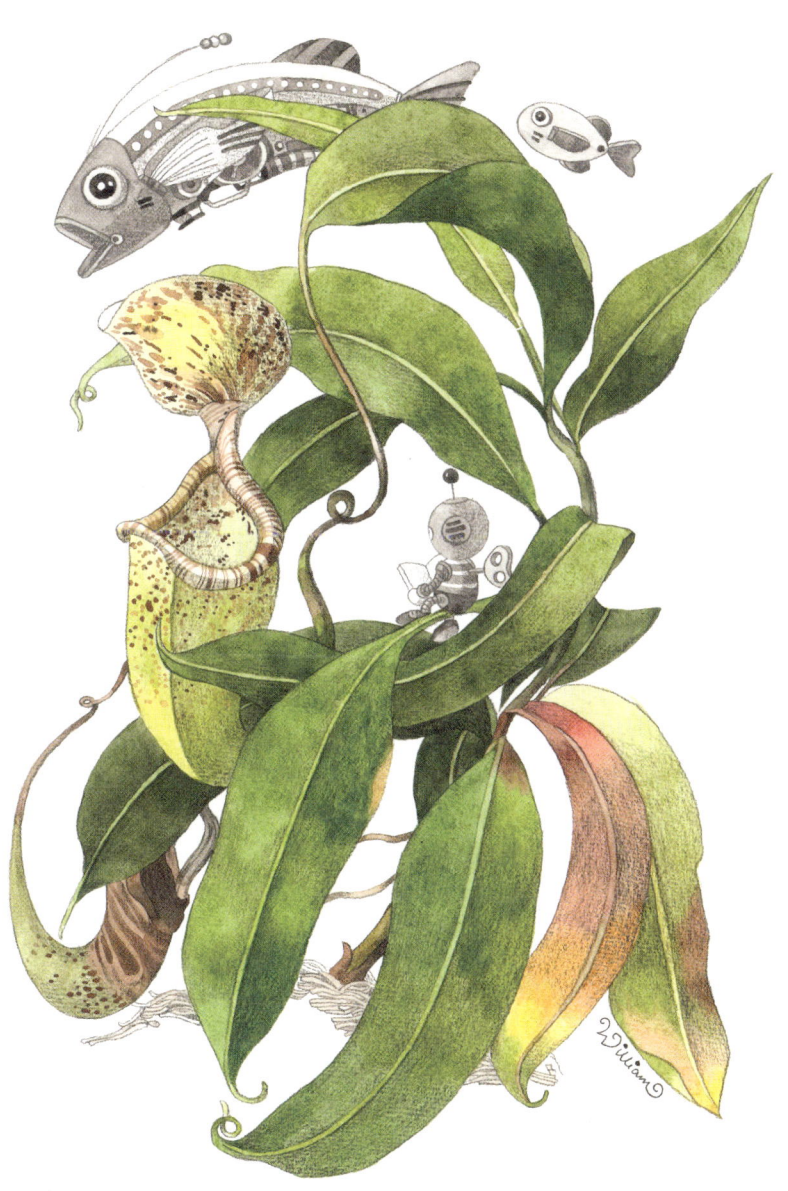

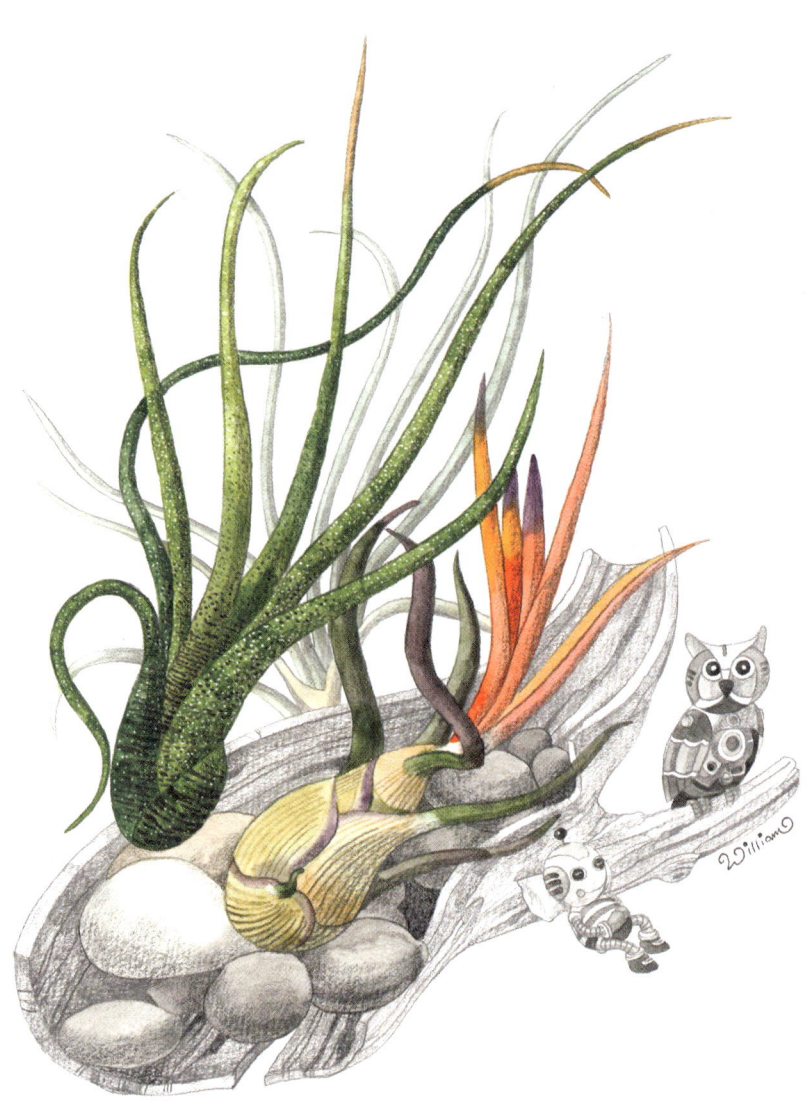

Tillandsia

BROMELIACEAE

Tillandsias are epiphytes and do not need soil as water and nutrients are absorbed through their leaves. The roots function mainly as anchors. These small airplants grow in dense clumps or rosettes. They have thick, narrow leaves with the innermost leaves (nearest to the centre of the rosette) turning bright red when the plants begin to flower. It is believed that this helps to attract bird pollinators such as hummingbirds and sunbirds.

Pomegranate
PUNICA GRANATUM

The pomegranate is a small bush or tree that grows up to 3 metres tall. It has long leaves that are shaped like a lance. The red flowers are found on the upper part of the stems. The fruits start as berries but become round leathery spheres when ripe; they are red-brown in colour and can measure up to 12 cm across. Inside, the fruit is divided into compartments filled with seeds encased in a juicy pulp. They can be eaten on their own, and are also mixed in salads.

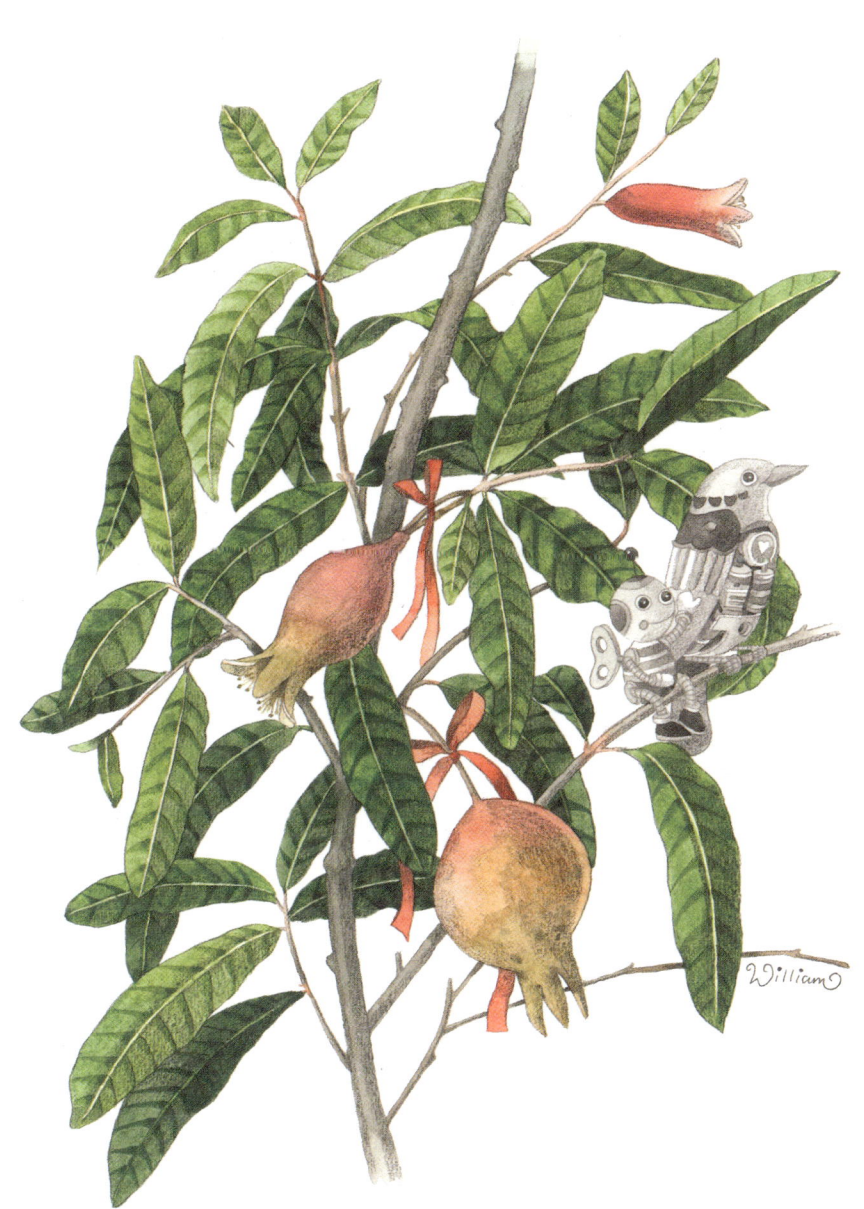

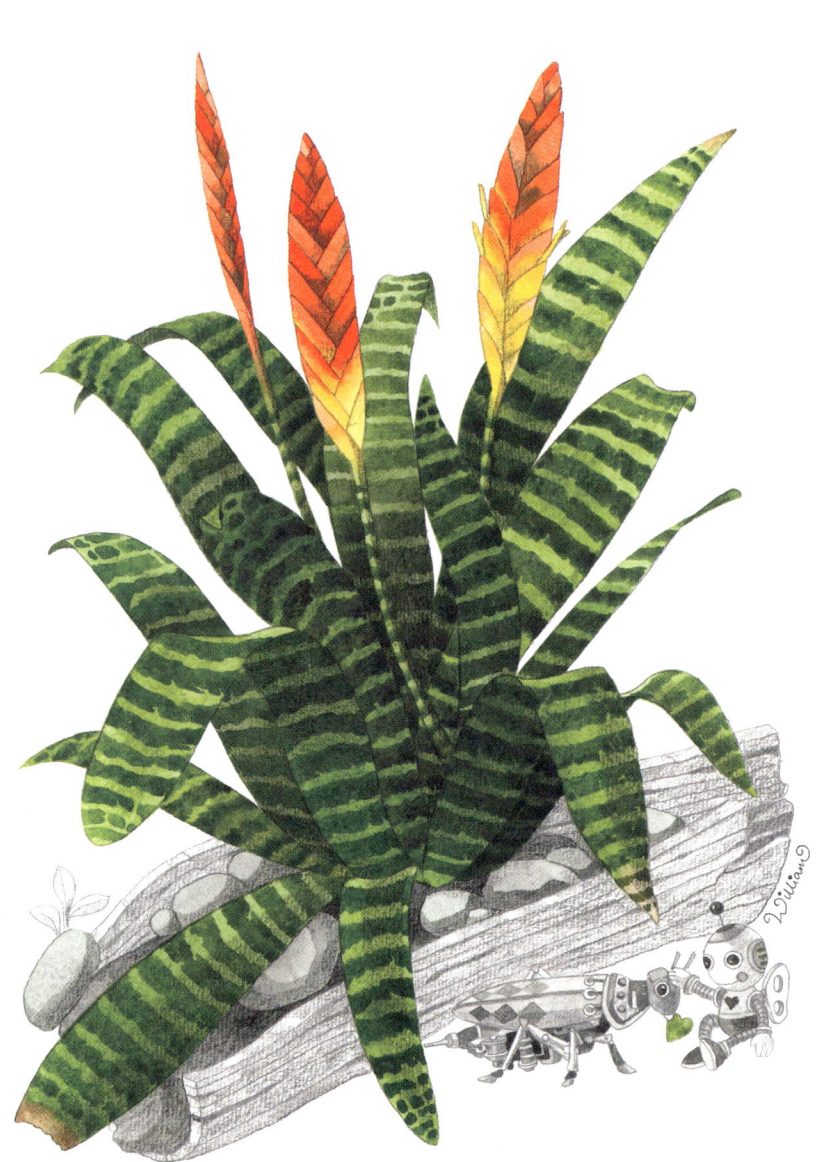

Vriesea Splendens
BROMELIACEAE

This tropical plant is another epiphyte. It grows on trees using special hold fasts that do not take in any nutrients from the host plant. All nutrients are taken in through the central "tank" made by the rosette of leaves. The leaves are soft and have markings along the upper and lower surface of the leaf blade. The yellow flowers and red bracts are arranged in a spike on a stem that can grow to a height of 45 cm.

Desert Rose
ADENIUM OBESUM

The Chinese believe that this is an auspicious plant as it has thick swollen stems that represent fertility and abundance. It also has different coloured flowers with the red and pink variety being more popular as these are believed to signify luck and prosperity. The plant has simple, glossy green leaves arranged in a spiral pattern. As it is easy to cultivate and can be grown in containers, some people also grow it bonsai style.

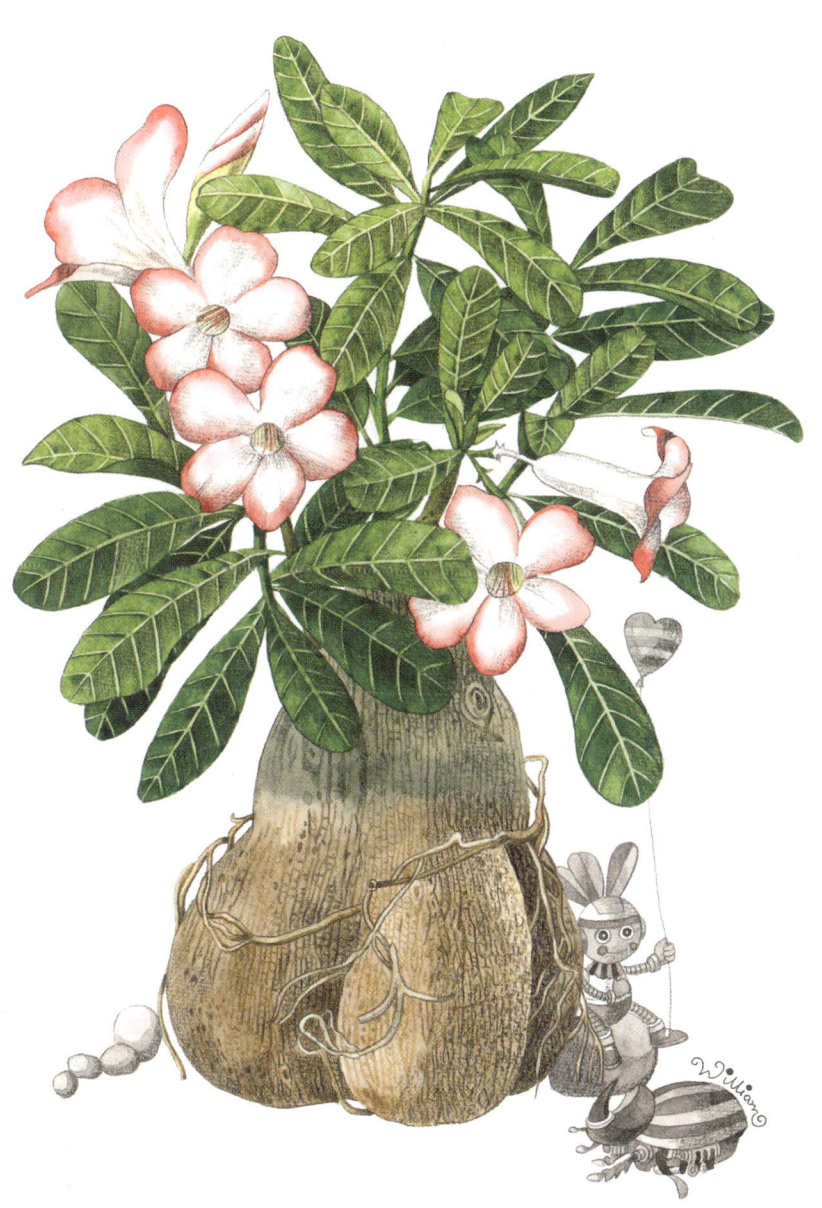

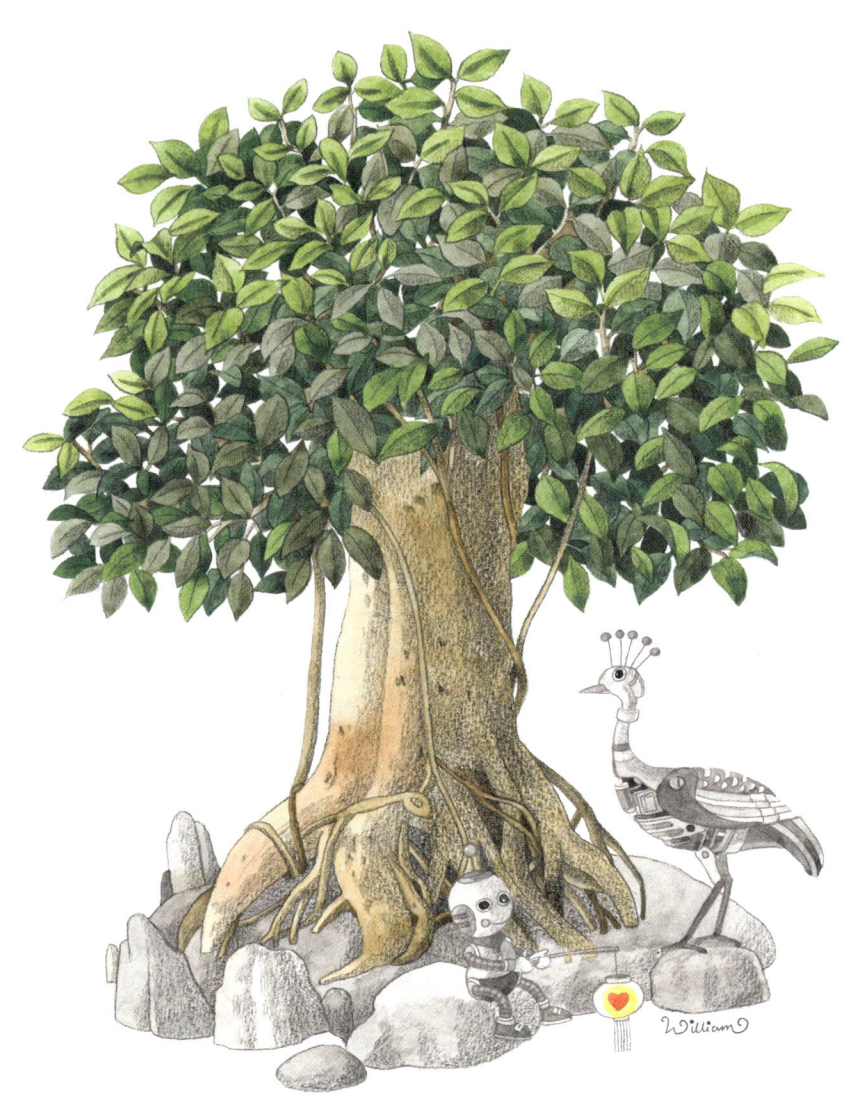

Ficus Benjamina
MORACEAE

The Ficus is also called the Weeping Fig as it has drooping branches with glossy leaves. It can grow to a height of 30 metres and is often found in parks and gardens. The fruits or figs are small and reddish-orange in colour. It can also grow indoors. A minature variety of the plant is also available and its shape makes it a popular choice for bonsai.

Calathea Medallion
MARANTACEAE

The Calathea is also called a zebra plant due to its colourful leaves that resemble the striped animal. It has broad, dark-purple leaves with a prominent silver central vein or mid-rib. It is often grown indoors for decorative purposes as they can survive in dim light as long as there is sufficient moisture in the soil.

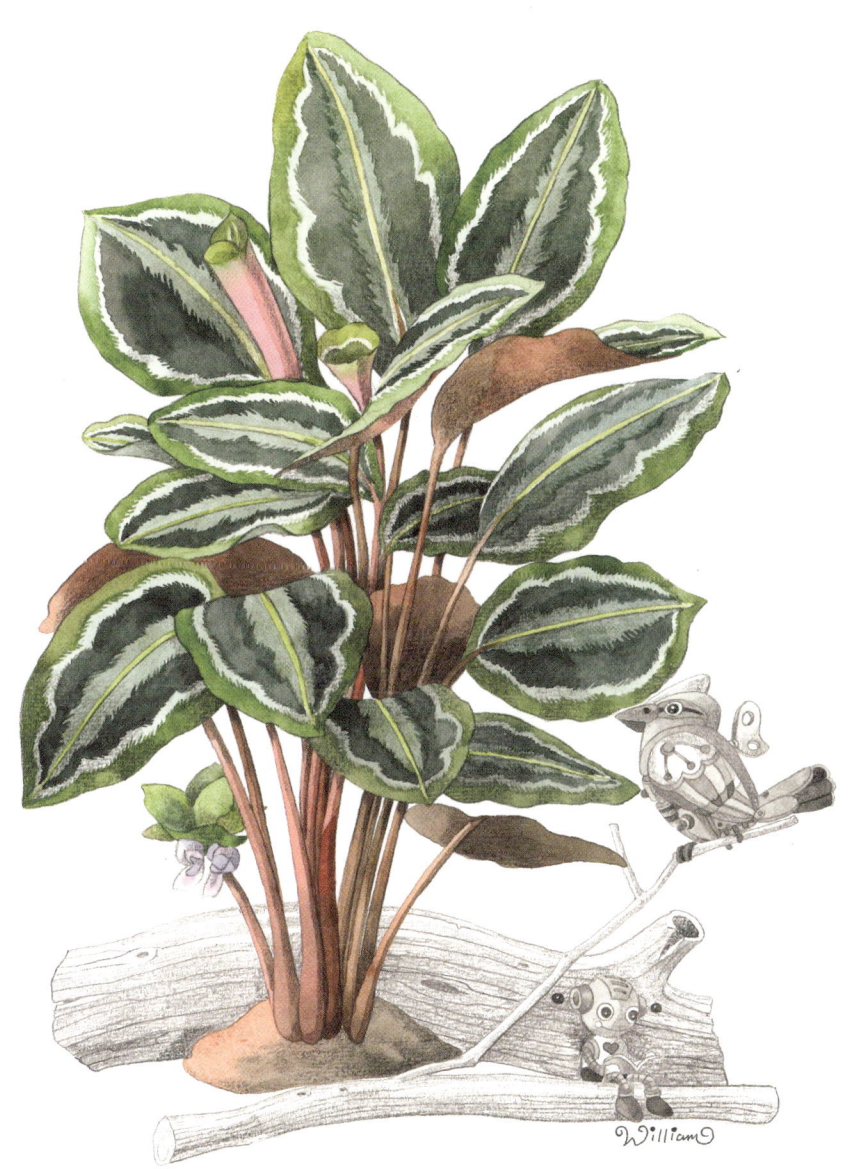

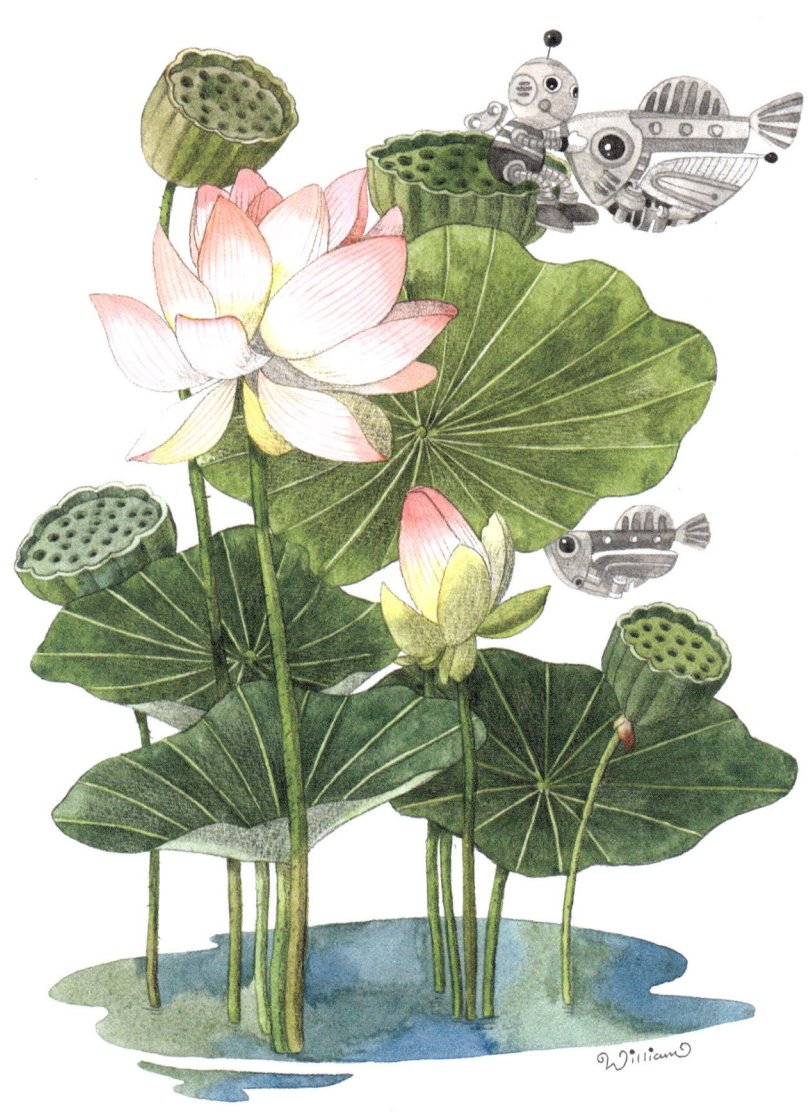

Lotus

NELUMBO NUCIFERA

The roots of the plant grow in the river or water bed. The leaves which can be as large as 60 cm in diameter float above the water along with the flowers. The Lotus has a long life and can live for hundreds or even thousands of years under the right conditions. The seed heads are very distinctive and dried seed heads are often used in floral arrangements. This is one of the few plants where the entire plant can be used in traditional medicine.

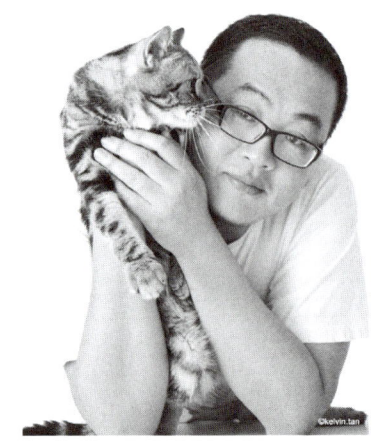

About the Author

William Sim's art often portray a fantasy world in a kaleidoscope of candy hues and shades. The subjects, a blend of nature and mechanical objects, collaborate individually and collectively to depict dreamscapes of immense optimism.

Branding himself as the Merchant of Happiness, William is a full-time art practitioner and a partner at a visual arts studio based in Singapore since 1997. He has showcased his paintings in various group and solo exhibitions in Singapore and countries such as South Korea, Taiwan and Hong Kong.

William Sim is also the illustrator of four adult colouring books: *Colouring the Lion City*, *Colouring the World*, *Colouring Chinoiserie* and *My Zodiac Colouring Book*